Library of Congress Control Number: 2017950209

Designed by Brenda McCallum
Cover design by Matt Goodman

Type set in Nexus Sans/ Nexus Serif/ Kleymissky

ISBN: 978-0-7643-5453-3
Printed in China

Published by Schiffer Publishing, Ltd.
4880 Lower Valley Road
Atglen, PA 19310
Phone: (610) 593-1777; Fax: (610) 593-2002
E-mail: Info@schifferbooks.com
www.schifferbooks.com

For our complete selection of fine books on this and related subjects, please visit our website at www.schifferbooks.com. You may also write for a free catalog.

Schiffer Publishing's titles are available at special discounts for bulk purchases for sales promotions or premiums. Special editions, including personalized covers, corporate imprints, and excerpts, can be created in large quantities for special needs. For more information, contact the publisher.

We are always looking for people to write books on new and related subjects. If you have an idea for a book, please contact us at proposals@schifferbooks.com.

Other Schiffer Books on Related Subjects:

Cape Cod and the Islands: Where Beauty and History Meet, by Kathryn Kleecamp, ISBN 978-0-7643-5305-5

Barns of Cape Cod, by Joan Dillon, ISBN 978-0-7643-2564-9

Birds of Cape Cod & the Islands, by Roger S. Everett, ISBN 978-0-7643-2461-1

This book is dedicated to our parents,
Chet & Selma and Parker & Helen, to our brother Brian,
and to Audrey, the wonder dog, with love and gratitude for the time
we shared on this Earth together.

CONTENTS

INTRODUCTION

Welcome to our blueberry farm. We hope our story will capture something of the essence of farming life and why it is so important to us. There is enough science to be serious, which we are, and enough creativity to be spirited, which we are as well. Although this book is deeply personal, it is not a memoir. We are in it, of course, and it is non-fiction, but center stage, under the limelight, the lead or main character is always the blueberry. We are in supporting roles. It is a chronicle of a composite year in the life of our farm, which is to say our lives as lived on the farm.

Our blueberry farm is on Cape Cod, a mile up from the ocean in the town of East Dennis, Massachusetts. The sandy peninsula we call home would not even exist were it not for the last glacial retreat 20,000 years ago. When we walk at the farm and the beach nearby, we sometimes try to visualize an ice sheet two miles high above our heads. It is hard to imagine, but that is exactly what was here before we were. When the Laurentide Ice Sheet retreated over the course of 8,000 years, it left behind this fragile coastal peninsula scattered with immense boulders. The most well-known, Doane Rock, is in the Cape Cod National Seashore in the town of Eastham. These stones are called glacial erratics and we named our farm, Hokum Rock, after one of them. Hokum Rock is the second largest glacial rock on Cape Cod and is down the street from our field. There is another enormous stone on the beach where we walk and one hidden in a local wood. We might very well call them the Dennis Standing Stones. Several smaller ones, but no less impressive, came to rest in what are now our woods. We think of them as touchstones to the past, and they remind us that our lives are brief in comparison. They are rocks of great power—remarkably beautiful, awe-inspiring natural sculptures.

Our land has been in the family since the early 1940s, and we started farming it in 1973. The farm is on high land, or a glacial moraine. When we first started farming, we raised vegetables and strawberries with our first blueberries. We dedicated the farm to blueberries exclusively in 1986. Blueberries have grown wild here since native vegetation began to flourish after the glacial retreat. In fact, the scoured, acidic, and sandy soil was perfect for them, and with enough water and abundant sunshine they flourished here. Some of these woody perennials can live a half-century or more.

Glacial erratic on East Dennis beach. | BELOW Glacial erratic at farm I.

Glacial erratic at farm II. | BELOW Glacial erratic known as Hokum Rock by the farm.

Some of the wild blueberries in our woods stand more than seven feet tall with gnarled branches twisted into fantastic shapes and painted with bright green lichen. Cultivated blueberries, such as those we grow, were not established on the Cape until sometime after 1920, when they were first introduced in the state. Of our present eighteen acres, six were originally cleared for the farm, with three put under cultivation; now we cultivate 1½ acres, and our blueberry plants number 1,500. It is a small farm by most standards, but when working on hands and knees, it seems pretty large. Despite its size, it requires a lot of work and produces thousands of pints of berries each year.

Our blueberry farm is not set in some idyllic rural area. It has no gorgeous view of mountains, valleys, rolling agricultural lands, or open water. It is on Cape Cod, a beautiful and popular tourist destination, known for its beaches and charming New England seaside villages. Over the last six decades, the number of larger farms that once existed here slowly decreased as development pressures and increasing land values changed much of the land use. Stores, businesses, and housing for locals and especially for summer residents, expanded at a rapid pace. In contrast to this development trend, we began a farm on our family land, last farmed by the previous owner forty years earlier.

Throughout this period of change, the towns, state, and federal government, which established Cape Cod National Seashore, periodically bought large tracts for open-space wildlife habitat, recreation, and water protection. Although our farm was carved out of ordinary woodland, it occupies its place in this diverse landscape. As the photographs illustrate, the farm has a special kind of beauty and offers a different type of open space, one that produces food for humans and wildlife. It was deliberately designed to create a sense of being isolated from buildings and roads. We left a boundary of trees to screen the roads that run along two sides. Deep woods border the other two sides. The overall experience for our patrons is one of being in a natural setting, which enriches their time here, and ours, as we work and walk around the farm. Those without gardens might gain an understanding of, and connection to, how food is raised. Farms provide this value by being part of the landscape in any community.

We hope to provide you, our readers, with a portrait of our farm throughout the year, as well as something of the spirit with which we farm. We hope to show our respect for the natural world and the gratitude we feel for our relationship with it. Our lives derive meaning, in large part, from this relationship. We are committed to the responsibility we bear to honor the natural world and to act as good stewards of this small piece of land and its biodiversity. These fields and woodland are an integrated ecosystem with our cultivated blueberries at their center, supporting an intricate web of life, a rich diversity of flora and fauna.

As coauthors, we bring an interesting balance of experience to the narrative. Stephen Spear has worked for the United States Department of Agriculture (USDA) for more than thirty years in an agency called the Natural Resources Conservation Service (NRCS). He is a conservation specialist. The purpose of his agency is to support farmers and local conservation efforts. He holds a BA in English from the University of Utah and a BS in plant and soil science from the University of Massachusetts, Amherst. He has been a farmer and gardener for more than forty years. Janice Riley is a professional writer, photographer, and web master. She worked in journalism for more than thirty years, starting out with a small newspaper. She then became a documentary editor, director, videographer, and writer with PBS, *Christian Science Monitor*, and independently as Riley Films. She has also worked in education, as well as museums and libraries on the Cape. She has a BA in anthropology from Wellesley College and has been gardening for more than thirty years, blueberry farming for fifteen years.

It is true what they say about a labor of love—that the work itself is its own reward. We hope the story of our farm, also a labor of love, is a testament to the farming tradition, one we are proud to be part of. Blueberries are the heart and soul of our life here at the farm. We hope we have told their story well and that our story celebrates one of life's pure and simpler joys—picking blueberries on a warm summer day. Blueberries under the blue sky are such ephemeral things. They are made for summer; they are its very fruit.

At the farm and in our lives we try to follow William Blake's sage advice, "To see a world in a grain of sand, and Heaven in a wild flower." As overly sentimental as this statement might seem on the surface, at its depths sits counsel as wise as any sought on any path: to "hold infinity in the palm of your hand, and eternity in an hour" (*Augeries of Innocence*).

PART ONE | The Year

JANUARY

In which the blueberries are dormant . . .

In any season, upon entering the field, there is always a sense of arrival and comfort in the open blueberry field, the close woods, and the sanctuary it affords. At times, in winter, the silence can be absolute, disturbed only by the rustle of dry grasses in the wind, the rattling of weathered branches in row upon row of blueberry plants, and the warning call of a bird, most often the crow, whose woods and open fields these are. An attribute essential to a farmer's life, or any life lived close to nature, is a sense of delight. As Robert Frost wrote in "Dust of Snow": "The way a crow, shook down on me, the dust of snow, from a hemlock tree, has given my heart, a change of mood, and saved some part, of a day I had rued."

Besides the American crow, house sparrows, mourning doves, and eastern bluebirds, among others, are our feathered winter residents. Under the cloak of dark we hear the deep notes of the great horned owl, though we have never seen him. From what we have read of red-tailed hawks, who figure in our story all year long, most migrate to warmer climates in the fall and return in spring, but we count at least one who winters over in our woods. Like the owl, and unlike our smaller residents, they are powerful birds of prey, soaring overhead with their screeching call, or sitting quietly on one of our posts, keeping a watchful eye to the ground. Far and away the most beautiful sight, though, is to see them scouting low through the trees, following the narrow, open swatches between them, as if in a magical woodland. For all our wildlife, winter is an arduous time that tests the limits of their endurance. Not so for our woody perennials, the blueberries, snuggled under their blanket of snow and mulch. They need these cold winter months to come fully back to life every spring.

There is a stark beauty in the stillness of a winter's day when all that is superfluous is stripped bare, stripped down to the essentials of survival. We are at the height of our winter rest here on the blueberry farm. We notice things we might have overlooked when the bushes were in full bloom and the wildflower meadow was a riot of movement and color. The silver-gray bark of the older canes, paper thin and curled here, knotted and whorled there, calls you closer in. When there is a blanket of white snow, the effect of the near luminous bark and red-tipped canes is stunning. And there are often tracks to follow in the white drifts, measurements to make along gloved hand and booted foot, to identify which creature passed that way and left its story in the snow. Like artifacts of summer, the Queen Anne's lace still stands tall, its once green stems bone dry, its creamy vault no more than colorless filigree. The switch grass, pale gold, the silvered cups

of milkweed leaning, the soft rye grass brittle, and tassels of goldenrod left behind, now dun-colored and brooding. These are winter's flowers.

A farmer's life is intimately connected to the natural world, and no less so for a mere berry farmer. Traditionally, it was the farmer who understood how to read the signs in the sky more than anyone else in the community, like a sailor might at sea. Whether watching for constellations by night or cloud formations by day, our knowledge of these natural signs is a skill best seasoned over time.

In January, the Earth's axis is tipped away from the sun, which accounts for our winter here in the Northeast, even though we are closer to the sun than at any time during our 365-day orbit. The sun traces a lower arc in the sky, so direct light is diffused over a larger area and daylight is of shorter duration. The sun's low angle in January is off about forty-five degrees from its summer height. Science tells us that it is our movement and position, not the sun's, which causes this apparent lower arc and its brief daily duration. It may seem paradoxical to terrestrial beings like ourselves, who determine things from our own perspective and not as they truly are. It would make more sense to us that we would feel the heat of the sun more if we were closer to it, but alas that is simply not the case. Our planet may be closer to the sun than it is in summer, but the sun's warmth is minimal. The Earth is spinning faster in its rotation due to the magnetic pull of the sun's immense gravitational force.

January's snows follow December.

All of these effects from the celestial sphere have direct bearing on us. We still read the signs of nature every week of every month. We can see the shadows of the plants and our netting posts, which travel from east to west in the rows, elongated by the low angle of the sun in winter, shortening in late spring through mid-summer. This change is quite noticeable over the seasons. The blueberry plants are their own sundials. Although we do begin to see a noticeable lengthening of daylight each afternoon, as the days lengthen after the winter solstice of December, by month's end we will gain about forty-five minutes of daylight.

Winter has its own beauties, and more specifically, its benefits for our blueberry plants that require their rest. These plants have been here for almost twenty years now, though the farm is much older, and of the 1,500 plants, we have lost only a handful. In their ordered rows they stand, thirty rows of fifty plants each. Six feet high, the Northern highbush blueberry, *Vaccinium corymbosum*, stand sentry, shoulder to shoulder, against the cold. In the wild, blueberry plants are not so orderly. Our extensive drip irrigation system accounts for this formality. The roots of the bushes extend twelve to eighteen inches down to draw sustenance and stability from the soil. The roots are deep enough to protect them from freezing, but even the deepest roots can succumb in a hard winter, when the frost line can reach down three feet. Fortunately, this is a rare event. Further protection is offered on the surface, where all thirty rows are covered by a few inches of thick organic mulch. When there is snow, it acts as an insulation layer as well. Thus the roots are protected as much as they can be. Not so for the bare branches, called canes. These are vulnerable to damage by harsh winds. Deciduous shrubs and trees drop their leaves every fall as an adaptation to cold climates. Not only do the leaves not absorb enough beneficial sunlight in the winter, they would be too delicate to survive and would transpire too much moisture away from the plant. Bare branches are a strategy that minimizes this vulnerability but does not eliminate it altogether. The most extreme cold and wind will strip moisture from the branches, making them brittle and prone to breakage.

In the winter months, the color of the canes is most striking. The young shoots and canes, either closest to the ground or at the uppermost part of each branch, are their deepest crimson red, in stark contrast to the older silver-gray lower trunks and branches. There is a symmetry to the blueberry plants in their neat rows, yet each has an individual presence, shaped through the years it has weathered and the pruning it receives each spring. Highbush blueberries in the Northeast require about 800 hours of what we refer to as chill time. The chilling requirement of a fruit-bearing bush is the minimum percentage of cold weather required before it will break dormancy. The botanical math is a rather inexact science, calculating the total accumulation of hours in which temperatures are thirty to forty degrees F. They do not need snow, nor do they need extreme cold. Dormancy begins after the leaves drop in late fall in reaction to changing light and temperature

Snow falling on bare winter canes. | BELOW Snow storm.

Blueberry plants after a winter storm. | BELOW Snow-laden blueberry plants.

conditions. The canes of the blueberry bush then turn red, usually by mid-November, deepening in December. This coloration lasts until early March. During dormancy, the leafless blueberry plant is no longer producing chlorophyll through the process of photosynthesis, which provides the energy required to grow and produce fruit. Sap is withdrawn from the branches. This energy conservation is a survival mechanism during the harsh winter. What small amount of liquid remains in the plant performs a chemical transformation at a molecular level, further protecting the plant with a sort of natural anti-freeze. The plants will remain dormant until their chill time is achieved, typically in early March, when an internal response to the change in light and rising temperature begins.

Dried winter aster. | Ice crystals on winter grass.

Blueberry plants after
a winter storm.

January 23, is on average the coldest day of the year, according to meteorologists. In fact, we often experience a thaw about the third or fourth week of the month. The temperature will rise by ten degrees for approximately one week, but, since temperatures are at their winter low, it will not break their dormancy. They won't even notice it. It is noticeable to us, however, and is always a welcome reminder of warmer days ahead.

The full moon in January is called the Full Wolf Moon, according to the *Farmer's Almanac*, which we keep on hand. Mid-winter is an unforgiving time, often a life or death struggle for all our wildlife. Wolves were native to the Cape when the Pilgrims arrived in the early seventeenth century. They were completely eradicated by the mid-eighteenth century. Wolves not only frightened the early settlers, they were direct competitors for limited resources, plus they raided their farms. Today the coywolf, or Eastern coyote, a hybrid between a wolf and a western coyote, has taken their place, easily expanding their population and range to fill that vacant niche as alpha, or apex, predator since the mid-1980s.

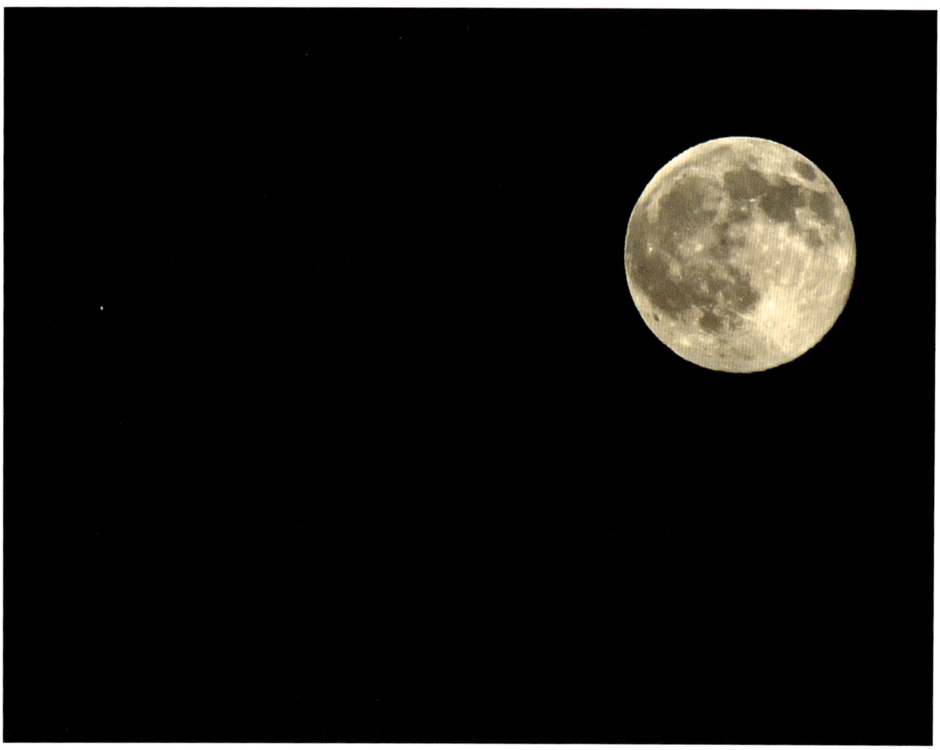

Full Wolf Moon.

Blueberry bark. | BELOW Horse-nettle berries in winter.

If one were to stand facing South at seven p.m. on any clear evening the sign of winter in the night sky would be unmistakable. Our winter constellation in the Northeast is Orion, that familiar hourglass shape that sits high in the January sky. It began its ascension in October and will disappear from view in March. The seasonal constellations appear to move across the night sky, when in fact, once again, we are the ones moving in our restless orbit. Even so, the shape of Orion rising in the night sky every fall, though a most beautiful star pattern, reminds us of time passing. It is almost as if the hourglass has been tipped and the sand is just beginning to accumulate, counting the winter hours in which nature's life and death dramas are played out. As a consoling thought, the stars at his belt, in the Orion Nebula, are known to be a star nursery, through images beautifully rendered by NASA's Hubble telescope.

Blueberry field in January.

Orion is actually called the Hunter, so named by the Ancient Greeks because its outline traced the shape of a hunter with his drawn bow. Orion will forever chase the summer constellation, Ursa Major, the bear, across the night sky, although he is also said to chase Pleiades, the unreachable seven sisters, who are a little closer to him. These beautiful stars appear even more brilliant on winter nights, as cold air does not hold as much moisture as warm air, which can obscure their radiance. There are also fewer stars competing for attention since the Northern hemisphere, in winter, faces a less populated area of the heavens. The Milky Way, for instance, the enormous swath of dusty light that is our galaxy, is not visible at this time of year. By mid- to late-January, Orion is no longer at the apex of his journey across the night sky. This change of position occurrs at nine p.m. in mid-December. As winter progresses, the Hunter appears farther and farther west of heaven's mid-point every night until he has moved entirely out of view by late spring. In his wake, the sparkling stars of summer's constellations trail.

FEBRUARY

In which shadows loom . . .

Maybe it's just wishful thinking, but our morning walks are accompanied by bird song now. Just a few, mind you, but a welcome sound nonetheless. And February is when the mourning dove breaks its winter silence. Although the flocks remain here all winter, we hear nothing more from them than their whistling wings in flight. The first time we hear their call, we realize how much we have missed that sound out in the field. After months of silence, there it is, those four gentle notes, a peaceful chord that lingers in the air. We stand absolutely still, with feet planted firmly on the snow-covered earth, listening. The notes repeat themselves and then there is the waiting silence. It is a mating call to spring.

To come upon tight-fisted buds at the tips of the blueberry canes, set last fall, seems another promise of spring on an overcast winter's day. In fact, February 2, Groundhog Day, was called Candlemas in earlier cultures and was actually considered the beginning of spring. A cross-quarter day, Candlemas falls halfway between the winter solstice on December 21, and the spring equinox on March 20. February, being that middle time, makes us pause and reflect. It reminds us that there are only six weeks, a very short time, until fair weather returns. It is said, in *Farmer's Alamanac* folklore, that "if on Candlemas it be shower and rain, winter is gone and will not come again."

Folklore also tells us that if the groundhog sees his shadow, there will be six more weeks of winter, and if he doesn't there will be less. In our experience there have never been less, and oft-times more. The science of the sun's winter angle does come into play however. Shadows are longer due to the sun's low angle, so it is more probable that they will be seen. It is said that more than the mere presence of his shadow, its elongated size, again due to the sun's angle, frightens the hibernating groundhog back into his burrow. This angle doesn't change from one February to another, but, if there is snow on the ground, it might appear more darkly menacing. One can imagine that the gigantic shadow a groundhog casts might be about the size of a fox or coyote, a very real predator, whose hovering form would certainly send him scurrying.

OPPOSITE Crimson blueberry canes contrast with snow. Crimson canes with snowfall. | Colorful canes with set buds.

Unfortunately, our farm's small groundhog population has disappeared. They never gave us much trouble, and we enjoyed seeing the shy creatures dashing for the safety of their burrows. Groundhogs are definitely pests in a garden but, oddly, they never damaged the blueberries. They kept their burrows along the wood line, and if we found an outlet among the rows, we simply filled it in. It's been many years since we've seen one, which coincides with the burgeoning population of coyotes; slow and fat, woodchucks are no match for such a swift predator. Much larger than a western coyote, though smaller than a wolf, our local coyotes are more than a little fearless. We are their only predators. Their plaintive cries at night, their yipping and howling, is something we have grown accustomed to, though not easily. Numerous family units, or packs, populate the Cape and are often seen cutting across paths and roadways. At the farm we have most likely lost our groundhogs to them and our once abundant rabbit colony as well, though their numbers fluctuate. A pack can take down a deer. One fall we found tuffs of fur and deer hide in the field, and the following spring, a deer skull deep in the woods.

Blueberry shadows.

OPPOSITE Moody bare canes after snowmelt on February day.

When we do catch a glimpse of a coyote, it is clear that this is truly a glimpse of the wild. It is magnificent and unsettling at once. We will long remember the night one of us was nearly knocked over by one on our way to the garden. The coyote was tearing past on a hard run, so close we could have touched him. We could see his narrowed eyes, lolling tongue, and gleaming white teeth. He was panting heavily, intent on some pursuit we were blind to in the dark. Though he must have sensed us, he did not turn. It is something we will never forget. Though sightings seem to be less and less frequent, we know they are there. They cannot remain so invisible after a fresh February snow when we can follow their straight, unwavering tracks out of our woods and into our field.

They say that coyote and fox never share the same territory, but that has not been the case on our farm. For a couple of years, a thin, juvenile coyote took an interest in our work, watching us intently, and occasionally running alongside the tractor like a dog. A resident red fox found us equally mesmerizing and often watched us from his perch atop the mulch pile. We took pleasure watching him hunt for field mice, with his remarkable leap and pounce, and the flounce of his fiery red tail. As wildlife goes, the rabbits and groundhogs have always seemed less wild to us, as prey rather than predator, more closely aligned to the dove than the hawk, but they are no less wild.

Long-distance tracks.

The long shadows cast by our netting posts and plants certainly point to more wintry days ahead. Our blueberry bushes cast intricate dark shadows of their own on the drifts of white snow, unlike any other time of year. All is quiet at the farm except for the familiar crunch of snow underfoot. The plants are still slumbering, their flaming crimson canes ablaze against the snow. February is the month during which we receive the most snow, though it varies from year to year. The full moon is appropriately called the Snow Moon. As if in a parallel universe, the blueberries sleep, unperturbed by the weather around them, pulled down deep within themselves, dreaming, perhaps.

Depending on the severity or mildness of the winter, there may be some stirring among our plants by month's end. March is when we traditionally start pruning. The blueberry plants will let us know when they are emerging from their dormancy by a change in color. The deep red canes will begin to lighten toward orange. The science of their sleep follows an inner clock that has its own numbers; yet, it is beneficial that there are gains in sunlight; its strength and duration grow with every passing day. If they have had even their minimal chill time, an alchemy will begin deep in their roots and inner chambers and their sap will quicken.

The winter months see a great variability in weather extremes, but cold, snow, and wind are givens here in the Northeast, even if no two winters are alike. One year may be fairly mild with light snow cover, another bringing heavy snows of two feet, or no snow to speak of but bone-chilling cold with sub-zero wind chills.

Blueberry row at sunset.

On the Cape we are susceptible to ocean-effect storms, or squalls, blowing in and out, painting the trees with ice. The field is necessarily wide open and flat. That way it receives the full measure of sun during growing season, seven continuous hours. Just so, the field is open to the winter winds. Many of our winter storms blow in off the sea from the Northeast. Extreme winds can gust more than fifty miles per hour, which is not unusual this month. We watch for breakage in January and February especially, though it is usually minimal.

The Eastern bluebird and the American robin are among our wintering birds, though most migrate south. Still, to see a flock of twelve bluebirds in the field on a February morning seems a herald of spring. It is rare to see so many together. We have read that smaller birds huddle together in thick evergreens for warmth and survival through some of the more brutal winter evenings. By day, they stand atop the field posts and gather along the netting lines, as birds do, on the lookout for any movement below, swooping down and back up on their startling blue wings. Robins are common enough, but it is only when spring is truly upon us that immense flocks arrive in the field. Now the flocks pecking in the snowmelt are only three or four strong, thin and anxious looking. Our fields and meadows are at least a reliable resource for them in all seasons with their dried grasses and seed heads. And there are lichen and moss, acorns and pinecones in the woods, where a Northern flicker can tap off flaking bark on an old pine to find nourishment beneath.

We are nearly an island here on this narrow strip of land, literally so, if unnaturally, due to the Cape Cod Canal, which separates us from the mainland of Massachusetts. We are surrounded by the sea. We have a pond for every day of the year, it is said, 365 reflective surfaces, all that hold light much longer than the land. And there is so much sand, composed of microscopic flakes of quartz crystals, refracting diamond-like light, as does the snow and ice. All this light is enlivening. On February days, when the clouds are low and gray and the field an expanse of muted color, the light is often incandescent. We never tire of seeing what it will bring into focus. The angle of incidence, that point where light hits a surface, making something visible, is never the same twice. There is beauty to be found in even the harshest times. By the end of February however, with its short days and long nights, the inhospitable frigidity of winter can wear thin and make us long for a more forgiving season.

MARCH

In which pruning begins . . .

Whenever we walk out into the field on a March morning, we can't help repeating words that seem to rise up unbidden from our memories: "March, that season of the whole air hesitating and blowing back and forth . . . It is waiting weather." These were written by our dear friend and mentor John Hay in 1958, in his beautiful book *The Run.* We know, as he did, that our hopes will rise and fall this month, with days of warmth and sunshine and others of relentless, chilling rain. We will be disappointed again and again, but spring is slowly inching forward, advancing and retreating, as we will do. We think also of the words of Greek poet George Seferis: "If I wished to stay by myself, I desired to find Solitude, I did not desire such endless waiting, the scattering of my soul to the horizon" (*Poems, IX*). Do we overdramatize the point? From what we know of physics, time moves more slowly at sea level, which is precisely where we are. We have not yet put winter behind us, but what is most palpable in the air now is a sense of calm purpose. Gone is that feeling of life's impermanence that shrouded our fall and winter days. Now the earth is stirring, slowly coming back to life, though still largely invisible to the eye. Some say there is no spring on Cape Cod. To us, spring is a long, drawn-out affair that extends over a period of three months, tantalizing us with its promises. It may be waiting weather, a time in which our waiting seems endless, but at the farm there are things that will not wait. It is also pruning weather.

Like ground-based semaphores, flags have been set on the first rows we will prune. We always start with the hardest rows. In the old days, when we were flush with the vigor of youth and its heady sense of independence and invincibility, we used to prune this entire field of 1,500 plants by ourselves. Some years, our friend Ed, a silversmith, lent a hand. It is an enormously daunting task now, which is why we have helpers, two men and two women. As we are rarely in the field together at any given time, we had to devise a practical system of communication. Our system is simple and color-coded. It consists of three flags: red, yellow, and blue. The blue flag marks the row needing large cuts of old canes. The yellow flag is for thinning new growth. The red flag signifies small detail work, which is the last phase of pruning. As a worker moves down a row from north to south, the flags are placed where the work last left off. When a row is done, all three flags are laid down on the north side.

Robin forages for worms. | BELOW Flags mark rows for pruning.

Pruning a blueberry bush seems deceptively easy. But if you are standing in front of a six-foot-tall bush with shears in hand, you have a lot of decisions to make. The goal is to provide the plant with an opportunity to grow freely, unhindered by its companions on either side. Branches must be judged for the number of buds set and the number of other branches that will rub against it or obscure its sunlight once it leafs out. Pruning must accomplish all this, as well as removing old canes that are unproductive, freeing up space for new growth from the ground. The pruner must judge the potential of each plant, though remaining uncertain of the way it will actually grow. Some plants are pruned heavily one year and lightly the next. Every decision leads to a cut that will change things and lead to the next choice or two. There are several decisions that could be the right one, but once done, the pruner must follow that particular pattern of pruning. Mistakes can be made. Some pruners might worry too much, some maybe not enough. Some might appear too slow, some too fast. It is a balancing act that takes experience. Eventually it becomes easier to make the call. One thing that distinguishes our farm from other, larger farms is the attention given to each and every blueberry bush every year.

Pruned cane.

ABOVE Steve pruning.
Swelling blueberry bud. | Wagon full of pruned branches in March snow.

Pruners gather. | BELOW Bob and Dave pruning.

Once, we had a pruner who stood before each bush in a meditative pose for some time before starting in. Cocking his head, circling the plant, bending a branch this way and that. Now this is something we all do, but generally not for half an hour. It took him an entire morning to prune three blueberry plants. Admittedly, he did a masterful job. He was somewhat of an artist, in the tradition of Michelangelo perhaps, chiseling away at his marble block, freeing angels. After a long arduous morning in the field, he told us, handing in his shears, that he found the decision-making process too stressful.

Some pruners become what we call master pruners. Ours are. They shadowed us for a season or two before mastering the science and art. Only one of our helpers, Dave, a jack of all trades, whom we fondly refer to as the manager, does what we call the big cuts, along with Steve. These cuts involve weighty decisions for the plant— important choices that consider the bigger picture, plus the electric shears are a lot heavier and so are the bigger branches. Dave and Bob, a retired chef, usually work together pruning one bush at a time, with Bob responsible for more intricate pruning. They get a head start of two weeks over the detail pruners who follow in their footsteps. Mary Jane, a retiree from Woods Hole Oceanographic Institute and substitute teacher, and Sandra, a retired gallery curator and landscaper, do the detail work. They catch up quickly. Out early in the field, they are usually one row behind the big cuts in short order, closing in hourly.

The team works without our direct supervision, but we determine the order of the pruning, check their work daily, and give them verbal feedback when we are not out there pruning ourselves. The comments go something like this: too much on this plant, too little on this one, leave more new growth here, these unproductive canes were missed there, see how that branch rubs against the other, this area should be opened up a bit more for sunlight—that sort of thing. Our workers are all early risers. Since our mornings are often taken up with other obligations, we are more often in the field later in the afternoon. If all goes well with the weather, it will generally take all of us six to eight weeks to finish the field. It's nice to have our help back in the field with us, bent over the plants, laughing and chatting with each other. We have been fortunate.

There can be some disagreements about pruning, even between ourselves and among our helpers. Though the intent is to stimulate growth and yield and taste by pruning twenty percent of the plant annually, there is a subjective side. There is no one way to accomplish the goal. One of us is a heavy, fearless pruner, the other tentative, subject to erring on the side of beauty. If your pruning is too light, there may be more berries, but they will be smaller and less sweet. Under-pruning can leave berries subject to overcrowding, with not enough circulating air and sunlight, so they retain moisture, which encourages disease. If the pruning is too heavy, there will be too little fruit. The goal is a balance between the two. Remembering that we are only pruning less than a quarter of the plant, deciding which canes should go and which ones stay, is based on experience with an eye to form and function.

Farmyard greenhouse, pump house, and barn.
BELOW Bluebird under a blue sky. | Water pump meter.

In early spring, the irrigation system has to be checked. Each plant has two micro-irrigation sprayers, which is a form of drip irrigation. Installing this system of 3,000 sprinkler heads and the hoses buried between them was tedious and meticulous work. We had help from Karma and Nadoop, two of the many Tibetan refugees who worked on the Cape under a guest worker program. They worked tirelessly and without complaint, their broad, ready smiles radiant as sparks of sunlight. The field always seemed more peaceful when they were around and for days after. The program is not as strong now, as green cards are not easy to qualify for. Many in the Tibetan community moved to New York or returned to Nepal where they lived in exile. We miss Karma and Nadoop and another helper, Chewang, along with Tsering, Chewang's wife, who worked with us for several years on loan, as it were, from our friend Tim's farm.

One year a coyote, we believe, dug up several spray heads and gnawed the plastic hoses as if they were chew toys. He did this more than once over the course of several days. Thirty spray heads and yards of line were damaged. We waited him out, and when he finally tired of the game we replaced them. It was much harder and far more expensive another year to evict the family of field mice that had taken up residence in the buried irrigation lines, although we admired their clever ingenuity. Even without these small spring trials, each nozzle head has to be thoroughly cleaned and all the lines and heads flushed with fresh water to unclog them. We do not have our own well at the farm. We use town water. Our watershed, the unofficial pump house, is attached to the back of one of the greenhouses. This is where all our irrigation efforts originate, using an above ground water line running from the pump house, through the woods, and into the buried irrigation system that sustains our fields.

The spring, or vernal equinox falls on the 20th of March when the days and nights are of equal length, marking the midway point between winter and summer. Some days there is a pungent scent of softened earth in the woods, a heady aroma carried on air that feels as clean and fresh as pure spring water. One day the sky is a warm brilliant blue scudded with white cumulous clouds, on another a dull ceiling of tousled gray rags, edges ripped and frayed. There is often rain. We are impatient and still waiting, but April lies ahead. Spring rain can be delicate and musical and nice to walk in—yin rain, as they say in the East—but a good soaking yang rain is good for everything growing now. It can be cold this month and it seems colder because our hopes have been raised. It can snow.

The color change the blueberry plants undergo as they emerge from dormancy is subtle. In fact, it is only apparent from a distance. Standing at the far, highest end of the wildflower meadow, looking back at the fields, the change in coloration is clearest to the eye. From a distance, the eye can take in the contrast with the pale yellow grasses, the bare gray branches, and the deep green of the pines. Interspersed, there are pathways and roadways, so well worn they are lined with sand, bleached white by the sun. When we stand deep in the rows, however, the bushes still appear winter red. The color is changing, though, growing paler, and this means the blueberry plants are responding to the stronger light and warmer temperatures, awakening from their long winter's dormancy right on schedule. The tender buds that will begin to swell on the branches this month will soon become flower and leaf. The color we are beginning to see on the blueberry canes is similar to a robin's orange-red breast. This is a natural comparison to make, as huge flocks of these emissaries of spring dash about beneath the blueberry plants, tossing aside leaf mulch in search of scurrying insects and burrowing worms. The full moon this month, aptly, is the Full Worm Moon. Worms aerate the soil, and their presence assures any farmer or gardener that the soil is rich and good for planting.

A pair of red-tail hawks circles overhead, immense and downy-white against the blue sky. Their swag of fiery tail feathers catches the light. They circle low and seem to call to us. Native American friends say they circle sacred space, and we like to think it's true. They seem to be looking at us, tipping their wings, repeating their wild call. It seems fitting on a spring day to see two, possibly mates. They circle upward and float off as we circle the field. We have mowed the wildflower meadow, as we do every spring, to reduce the second-growth progression of woody shrubs and trees, and to promote spring growth among the herbaceous perennials. The ground is thick with stubble that hurts the dogs' feet, and they implore us to carry them. It is good to be out here. It invigorates the spirit and our work-tired limbs. This meadow will reawaken in the coming weeks, and so too will the hardy buds of the blueberry plants, holding fast now, awaiting warmer days, their color the lick of a pale orange flame soon to take fire.

Opening blueberry bud. | Netting post knot
BELOW Blueberry field on a March day.

APRIL

In which blueberry buds open . . .

The cold northeast wind of last month has abated and the sun is warmer. Out in the field among the rows of blueberries, as if in a sheltered valley, the warmth is held, the tree line a buffer. But in the wide, open meadow the wind still has a bite to it. The pruning continues, the flags moving up and down the rows. April is a transformative time here at the farm; the very air feels recharged with energy. That there are mysterious goings-on in the interior lives of the blueberry plants becomes ever more apparent. As we move among the plants there is a lively flexibility to the branches, and winter's crimson canes have paled to spring's soft colors of amber, green, and gold. The tiny buds are expanding in response to the light and warmth, and life-sustaining sap courses up from its active root system.

This long-awaited month is when our farm work begins in earnest. The weather is far more favorable now for steady pruning, and weeding has been added to our ever-growing list of spring chores. Starflowers and mayflowers are blooming safely in the woods, bright sparks on shadowed pathways, but creeping through our rows are their nasty cousins, white henbit and purple dead nettle, making nightly advances. But the middle ground, between earth and sky, is taken up with the daily metamorphosis of the blueberry plants. By month's end the entire field will be an exuberant haze of spring's yellow-green. The blueberry fruit buds, once closed tight, open layer upon layer, straining their reach forth from the ends of the branches with rose-tinted tips. Tiny leaves emerge along the branches, in rear guard, like so many miniature velvet horns. The blueberry buds are small green torches with rose-edged filigree, no larger at first than the nub of an eraser at the tip of a pencil. Reaching sunward, the buds lengthen into tiny sheaths, whose flattened layers begin to separate and unfold, like so many stiff petals. This does not happen all at once across the field, but each plant and each part of the plant develops on its own inner timetable, with various stages of development present from bush to bush. Smaller, pointed nubs that appear along the stems will become leaflets, which unfurl within days of appearing. Each seems to shake itself loose, and as it does it is transformed into a delicate open palm made expressly for capturing sunlight on its nearly translucent skin. There are bees about their own business; there is little for them to forage for except the low weeds, where the sky-blue spring azure butterfly flits. It is the month of the Full Pink Moon.

OPPOSITE Spring light on blueberry bushes. | Swelling buds.

The osprey have returned. This is the largest raptor in our area, except for occasional sightings of bald eagles. We look for them every day above the farm as we prune the rows. There have been three nesting pairs in close proximity for several years. One nest sits atop the police station radar tower, one on a nesting pole created for them at the harbor, another in the center of a nearby marsh. They never alight in the field, being fish lovers, but often fly low in reconnaissance or circle in the high thermals, in spring a pair and in late summer with offspring.

Pruning is all we think about. If the weather has been fair, the temperatures can soar into the upper sixties and buds start to pop open everywhere. A chill of fear grips us then as we survey the rows we have to get to before the leaves and buds fully open. We need to stay ahead of them. We will lose days if it rains. We do not prune in the rain, not because we are averse to getting wet, but because pruning moist branches could spread potential diseases present among them.

OPPOSITE Swelling buds. | Field looking northeast.
Turkey promenade.

Turkey with
blueberry buds.

Larry's samples for lab.
BELOW Small pruning shears. | Irrigation nozzle.
OPPOSITE Pile of pruned blueberry branches.

Fieldwork works more muscles than you imagine and they are sore at the end of the day. Our plants cover one-and-a-half acres, or 65,340 square feet. We have calculated that walking up and down the rows from one end of the field to the other means we cover 2.4 miles in a round trip. Pruning requires bending, squatting, stretching, twisting, and cutting and pulling branches. Our electric pruning shears for the big cuts, and small handheld shears for detail work, require a strong, steady grip. We prune one quarter of the old-growth canes back annually on every bush. Cuts are made at an angle with a downward thrust that requires a good eye and precision. We do not leave any clippings on the ground, again to protect against spreading any potential diseases, although we do pile larger cuts temporarily before hauling them to the wagon. Smaller cuttings are dropped into plastic barrels we pull along as we prune. These fill up quickly and can get heavy. We haul or carry or drag them over to the old wooden wagon we leave parked at the end of the rows. At the end of the day, when the wagon is overflowing with branches, it is pulled by tractor to the mulch area to be piled up and left for burning in late spring. All of this is hard work. It is manual labor, but it feels like an honest day's work and our souls are generally satisfied. Pruning is also meditative work that keeps you focused on the present moment and the task at hand, a practice deepened by being out in the fresh, open air, warmed by the spring sun. Even the shears, with their metallic chirring, call a wandering mind back to its purpose, certainly if you happen to snip your fingers.

Our conversations in the field often turn to insects now. As we adapt to a changing climate and widespread distribution of alien insects and plants, new invasive insects are of particular concern here at the farm and require us to take protective measures. The winter moth and the spotted wing drosophila fruit fly are annual challenges, the gypsy moth only on occasion, and recently the brown marmorated stink bug. Bugs are on our minds and we watch for telltale signs of their presence. Over the season, we watch for the arrival, or absence, of many others—native bees, butterflies, moths, and fireflies, whose populations diminish by the year. We always keep an eye out for ticks. We learn new things about insects, both fascinating and frightening, from our entomologist friend Larry Dapsis. The

English plantain weed. | Wild garlic in rows.
BELOW Blueberry flower bud and leaves expanding.

farm works with Cape Cod Cooperative Extension, a government organization that was established in the early twentieth century. This wonderful agency provides educational assistance, research, and resources to farmers in every state across the country. An entomologist in each county works closely with local farmers and becomes a regular visitor at all the farms. This month and next, every other week, Larry will collect buds to bring back to his lab. There he will examine them under a microscope looking for early signs of infestation, starting with winter moth larvae in the early spring. He will report his findings back to us. Our farm is one of many he collects data from for regional statistics. He also sets monitoring traps all season, from spring to fall, to study the spread of invasives like the winter moth.

Introduced to North America in the 1950s, from Europe, the winter moth, *Operophtera brumata*, did not become a problem in the Northeast, or at our farm, until 2005. This nocturnal moth is unusual in that it appears in winter, usually early December, when males can be seen fluttering around outdoor lights in search of a mate. The mating season lasts from late November through January and the resultant eggs are buried under the surface of tree and shrub bark, including blueberry shrubs. The eggs will hatch out in April. They appear when forsythia flowers. Tiny green caterpillars emerge and disperse by a method known as ballooning. Each will set a guideline made of silk, similar to a spider's, and float from branch to branch or tree to tree, seeking nourishment. Their hatching is timed to coincide with spring's first buds and flowers, which they will eat voraciously. When the caterpillar is done feeding it drops to the ground, entering its pupal stage, from which it will re-emerge next November. Serious defoliation has occurred in our area, especially deciduous trees such as oaks.

The winter moth arrived on our farm in 2005, and decimated ninety percent of our crop. We could not open our gates to the public that year, only sharing what few berries there were with friends. It was a devastating loss. The following year the entomologist alerted us of the caterpillar's presence at the end of April. If we had not sprayed at that time, we might have lost our blueberry crop again, and if there had been complete defoliation, we might have lost the entire farm over the course of the next several years. Since that time, this has become standard operating procedure—we must spray once in April, possibly a second time in May, if they are present. We use an insecticide with the lowest possible toxicity so as not to harm other beneficial insects, birds, or animals. There is no method that doesn't involve spraying to control the winter moth, and, unfortunately, organic sprays are not as effective. Predatory flies are being experimented with for future control.

Some farmers treat their fields as a precaution even if there are no signs of infestation. We are more conservative and will wait until there is microscopic evidence of the moth, and only then will we treat the field if necessary. We are licensed pesticide users and practice every precaution whenever an application must be made. It is a meticulous process, not to be undertaken lightly, and then only on an early windless evening to minimize contact with wildlife.

Of course, we wouldn't have a farm without the help of all the beneficial insects out there—the ladybugs, the bees that pollinate, the worms that aerate the soil, or the beetles and grubs that break down detritus. Insects provide food for our birds, and the birds keep the insect population in balance. When it comes to ticks, we agree with poet Mark Doty who calls them "pinhead demons" (*Deep Lane*). Tick-borne diseases have become a real problem for humans and animals on the Cape and Islands in recent years, though domestic cats do not seem to be effected. We have read that the reduction in certain bird populations, extinctions caused by various, usually human-related, factors, has led to their increased number. There is a healthy deer population here as well, and we have been advised that wild turkeys are also hosts. The tick population fluctuates in three-year cycles, or so we are told, with more on the loose in spring and fall, though they do not actually die off in winter. Larry advises us to be educated about them, protect ourselves from them, but not to let them keep us indoors. Our rows harbor few ticks, but the areas of long grass are riskier. One of our workers fled to the nearest fire department one day where a fireman was kind enough to remove one from her scalp. She has a terror of ticks, but they do not keep her out of the field. Nor do the coyotes. She keeps a boat horn by her side in the event that she is ever cornered in a row, and we sometimes hear its blast from the farmhouse; it is deafening up close.

Another service of the extension office we benefit from is Buy Fresh Buy Local (BFBL). This educational initiative promotes the relationship between local farmers and the public. We catch up with them every April when we pick up our new BFBL farm guide, bumper stickers, and magnets at the Meet Your Local Farmers Day sponsored by the Cape Cod Farm Bureau and a local conservation trust. We always enjoy being part of this event. The Cape's farming community consists of 476 farms, though only about fifty take part every year. It gives us a chance to meet up with old friends and make new ones in the off-season when we have time for socializing. Each year, the farm table displays seem to become more inventive. We fashion a nest of blueberry branches on

Blueberry flower bud opening with unfurling leaves.

ours nestled with photographs of the farm, tips for growers, and recipes. A year doesn't go by when we aren't surprised by people's questions, and our pruning shears are always at the ready to demonstrate technique. Our favorite displays are the ones that include animals, and every year it's a bit different, from rabbits and roosters, to pigs, sheep, and once a shaggy, longhorn steer—could we bring farm dogs, we often wonder?

April is the cruelest month, or so wrote poet T. S. Eliot, "breeding Lilacs out of the dead land, mixing Memory and desire ("The Waste Land")." And Edna St. Vincent Millay said of this month: "It is not enough that yearly, down this hill, April Comes like an idiot, babbling and strewing flowers ("April")." April is National Poetry Month, and on our farm website we do not let it pass without notice. Nature and poetry are invariably intertwined. In our experience, though, April makes promises it will not keep; even if it is a month that raises our hopes only to dash them again, we would have it no other way. It is a trickster month and we are sometimes its fools. Eventually spring will gather the scattered pieces of any soul and rebind them. In the words of e.e.cummings, "Spring is like a perhaps hand (which comes carefully out of Nowhere) arranging a window into which people look . . . changing everything carefully . . . (Spring is like a perhaps hand").

In the hour before dusk, the sun has fallen below the tree line in the west, casting its long shadow over the rows of blueberries. That part of the field beyond its shadowy reach holds the soft glow of late afternoon light. It is the mirror opposite of morning's light. Walking on the wooded path back to the farmhouse, towed by eager, hungry dogs, we hear the creaking swing voice of the red-winged blackbirds. Often the first migrant to return in March, they are everywhere in April. And we hear the cheerful, sleigh bell song of spring peepers in the abandoned cranberry bog nearby, now a vernal pool made from spring rain, snowmelt, and a rising water table. It will evaporate by summer. Their ecstatic chorus grows in number and volume nightly. Overhead, pale lichen, old man's beard, has braided itself along the bare branches of the oaks and maples. It lends the woods an otherworldly glow, draping the tall trees, almost luminescent at this hour—it signifies clean, pure air.

MAY

In which blueberry flowers bloom . . .

May is a month of promises, made and kept. It is still spring, more fully spring, in fact, than it has been so far, and summer is a certainty that lies ahead. This month and next are ones we would all choose to linger in a bit longer if we could. The blueberry field is bright with spring's new green, but as the month progresses it will become a sea of snow-white flowers. "Look at the flowers," poet Rainer Maria Rilke tells us, "so faithful to what is earthly . . . eternal childhood fills them with grace . . . all those silent companions in the wind of the meadows" (*The Sonnets to Orpheus*, II,14). The warm days will fill the air with their subtle, singular fragrance—a tincture of lavender and lily of the valley, though milder, with a hint of honeyed earth. There will be times, standing in the center of a row, that it will sound as if you are standing at the center of a hive. The air vibrates with the bees' industry and they will literally bump into you as they hurry from bush to bush, then lift off in their straight flight to their hidden hives. May is like no other time at the farm, and it is a time that few people get to experience as we do.

Blossoming branch in afternoon light.

This month, among hundreds of leaves, all the blueberry buds will open into flowers and some will show the first small berries forming beneath. The blueberries' growth in these four short weeks is exponential and rather extraordinary to watch as they make their sure and steady evolution from dormancy into this fully awakened state. Theirs has been a measured and consistent growth under spring's fluctuating conditions; they are hardy plants. Our fieldwork keeps us busier than ever at the farm in May. Though we have grown rather tired of it, as a subject of thought and daily chore, pruning is ongoing. If the weather has been uncooperative, we are occasionally pruning late into the month, though rarely so. Our goal now is to finish as early as possible, and having that goal keeps us focused. So far, we have made great progress. Although it is customary and necessary to open up the plants for new growth from below, we are conservative some years and more liberal in others. There is always a balance to be maintained between what the plant requires for its vitality, and longevity, and a bountiful summer harvest. We want as many berries as possible, and yet if there are too many canes and leaves and berries on a shrub, all jostling for position, they will shade each other out. This will result in a smaller yield overall, or an abundance of smaller, tart berries.

Since the larger canes have all been cut during the first stage of pruning, our attention is drawn this month to finishing up our detail pruning. Not all farms have the time or patience for this crucial step. Detail pruning is also a learned skill and somewhat of an art, requiring that self-same steadiness of hand and keenness of eye. Decisions about where to cut still must be made, but the essence of the task is of a far different order—the plants, basically, are tidied up. All fruit spurs from the previous year that are now lifeless and worn are nipped off. Branches that cross one another will be pruned and congested interior branches thinned. Now that leaves and flowers and bees abound we have a better idea of this year's growth, but it can make our final pruning days more of a challenge.

Close pruning like this is a good opportunity for an intimate look at each plant. It is also an opportunity for a sense of relationship with the plants to take effect. Quiet conversations, not only between two pruners but also between pruner and plant often take place now; in fact, blueberries make good listeners. Talking to the plants is an acceptable practice in our field. They are so beautiful and so alive, it is hard not to comment aloud on their appearance and urge them on. We can feel the warm sun on our backs as we work. We can feel the breeze cooling the perspiration on our brow. We remind ourselves to look up every so often to take in the day or simply watch a fat bumblebee's progress from flower to flower, her head buried in the tiny white bell, with golden pollen satchels dangling from her skinny legs.

OPPOSITE Stigma of flower is clear within bell-shaped opening.
BELOW Bumblebees arrive to pollinate flowers.

Blueberry flowers in abundance.

Beltane, or May Day, is traditionally celebrated on the first day of the month with a maypole and a bonfire. We have a bonfire of our own at the end of the month when all the pruned canes are burned in an enormous fire at the far west end of the field. Beltane is another of those cross-quarter days, this one being halfway between the vernal equinox of March 21, and the summer solstice of June 21. This day marked the beginning of summer in agricultural communities of the Old World and many rituals were performed to insure the fertility and protection of the growing season. Household fires were relit from the Beltane bonfire and its ashes were used for blessings. Farmers and their cows often circled and even leapt over the flames. The bonfire was a symbolic gesture to the fiery sun, an example of what anthropologist James Frazer called "sympathetic magic," bringing new life to the fields from the ashes of the past. Our bonfire may occur at the end of May, not the beginning, and may be more utilitarian than ritualistic; still, we have our own hopes for a fertile and healthy season. May the fire's magic do its work in our fields.

The sun has climbed noticeably higher in the sky, rising from a more northerly location than due east, so it traces a longer arc in the sky. Since the equinox in late March, its light is much stronger, being more direct and of longer duration as it nears the summer solstice. This is all to the good for our growing season. In the night sky this month we see the Full Flower Moon, heralding our own flowers beneath, and the long-awaited return of the summer constellation, Ursa Major, the Great Bear, with the familiar stars of the Big Dipper. We have gained more than five hours of sunlight since January. Under the influence of the warming sun, the scent of millions of blueberry flowers now begins to fill the air, attracting pollinators. Blueberry flowers are bell shaped, similar in size, shape, and color to a lily of the valley, though fuller in the rounded bell and more iridescent than opaque. As the flowers develop, if you were to look closely, the tiny pearl-white flowerlets are tinted in places with shades of rose and lavender. The delicately colored flower clusters of cultivated blueberries are significantly larger and more numerous than their wild cousins in the woods, and they develop more slowly. But, regardless of where they are found, along pond shores and deep in every wood on Cape Cod, all blueberries are in bloom come May.

OPPOSITE Blueberry forming beneath withering blossom.
Cluster of berries, with withering and full blossoms.

Bonfire to burn pruned canes. | The fire burns slowly over several hours.
OPPOSITE Blueberry canes being added to bonfire.

To return to the subject of bees, for the past few years, as our plants grew in stature and the number of flowers increased, we have had to purchase bees to insure full pollination. The diminished native population of bumblebees and honeybees concerned us. Bumblebees are the best pollinators for blueberries and there just weren't enough of them around. Bees have been in decline worldwide for several decades, and we finally felt the effect. "If bees were to disappear from the globe," a quote attributed to Albert Einstein warns, "mankind would have only four years left to live." Much has been written about the causes of colony collapse disorder in honeybees and human intervention has played its part. We are not experts. We do know that they are considered a keystone species; in other words, our ecosystem would collapse without bees. According to the USDA, a full eighty percent of crop pollination is carried out by bees. Without bees there would be no blueberries on our farm. Native bees do visit our fields and there are many varieties of them, but there aren't enough. We know several beekeepers and considered setting aside a corner out back for transient hives, but honeybees were not the best pollinators for our plants. About 300 bumblebees arrive via UPS in mid-May. The beehive is a box with plastic mesh openings they gnaw their way through and will come and go from. The ultraviolet bloom on flowers, invisible to the human eye, will lead them to their destination. They will live here until they die by summer's end, as do most populations of wild worker bees. Wild queens survive until next spring, overwintering in the ground; ours may or may not leave the box to do the same. Our purchased bees look like any other bumblebees; in fact, they are, it's just that we've had to purchase them from a provider.

Northern flicker on netting post. | Bluebird on netting post.
BELOW Home sweet home: bird's nest in row of early green blueberries.

Bees are the agents of reproduction, busy cupids carrying pollen from one stationary flower to another. Bumblebees (*Bombus impatiens* are the ones we import) are considered the best pollinators for blueberries, although other species, even wild honeybees, also pollinate our fields. When a bee sticks its head into a flower opening (corolla tube), which is just the right size for it, and flutters its wings, buzzing as it hovers in place, pollen dust is released from the flower's stigma. The pollen dust falls and settles on the ovary, which contains the berries' seeds, at the base of the flower's bell. This is called buzz pollination. Pollen that dusts the bee's body and legs can also be transferred to the next flower it visits. Nectar, high-energy sugar water, is also provided to the bee in return for their efforts. It takes two or three days for the fertilization to be complete and berries to form. This is the most efficient method for getting the pollen from the tip of the stamen to the deeply hidden ovary and produces the best-quality berry. Bees gather pollen for their hives in little golden saddlebags on their legs. It is delivered to the hive for food storage and feeding the young. In contrast to the blueberry shrub, trees and grasses use broadcast pollination, which disperses pollen into the air, falling where it will, over the course of several weeks. Pine pollen will coat every surface with a fine golden dust all month.

Where there are bugs, there are birds. Although we may not be as enamored with insects as entomologists are, we do take great pleasure in the avian predators they attract. The hatching of insects in the warming field brings recently migrated songbirds and swallows back, their exuberant songs building to a crescendo during mating season, before the quiet of nesting and raising hatchlings descends. The bluebirds seem to team up with the other small birds—sparrows, chickadees, and finches—using our netting posts and lines as high sentries from which to launch their assaults. We have read that territorial attacks are common among these small birds, but we have never seen it. They sweep and dive with as much agility as the acrobatic, blue-backed barn swallows. A golden-backed and black-speckled woodpecker, the flicker, often joins them, though his flights are as direct as an arrow. Small flocks of doves, for all appearances, seem to use the same posts and lines for leisurely gatherings. Beneath, we have no lack of robins dashing about, cocking their heads from side to side over rustlings beneath the earth. They spend more time on foot than wing. More than once we have found robin and sparrow nests hidden among the blueberry branches once they leaf out fully, and the flickers raise a family high up in the same hollowed oak, with a perfectly round opening, every year.

As detail pruning tapers off, we attend to the rows at ground level with light fertilizer. In order to ensure its widespread effectiveness, all 3,000 drip irrigation nozzles must be exposed and checked. Those that are plugged must be taken apart and thoroughly cleaned. The irrigation lines will be checked for leaks and repaired or replaced if necessary, cleaned also, if clogged, and flushed with water to assure that the system is free flowing. We use two kinds of fertilizer: first, the mulch applied in the fall provides organic nutrients as it decomposes; second, we inject

synthetic fertilizer into the irrigation system once a week, for four weeks, as soon as the plants start to flower. By applying a little at a time over several weeks, rather than all at once, we reduce the chance that heavy rains will leach the nutrients below the root zone. We generally begin watering the field this month if rainfall is scarce. Weeding, once begun, continues unabated.

Clusters of wild violets are appearing in the meadow, with their deep purple and snow-white petals, and along the netting posts are swatches of pretty red sorrel, yellow hawkweed, and red clover. The sly and ubiquitous henbit and dead nettle weeds try to hide among them, muscling in, but not for long. And among the rejuvenated wild grasses bending and swaying in the warm breeze, we find nodding spike rush, twining grape hyacinth, diminutive dwarf cinquefoil, and the glossy buttercup. Every May, in the second row of blueberries, we seek out and find the strange and quite poisonous star-of-Bethlehem opening in the morning sun, for only this month, and a row beyond, our perennial spiderwort, which will bloom until fall.

Blueberry bushes in full flower.

As if to mirror the arrival of Ursa Major overhead, May is a month we will remember for the black bear that arrived on Memorial Day weekend in 2012. He was a young, wild bear and on the loose here for more than two weeks. Everyone was giddy with pleasure over the holiday vagabond and hoped for a sighting of the 200-pound, two-year-old juvenile. The last time a bear had been seen on Cape Cod was in 1837. It seemed everyone had a story of how he had crossed their yard or street, or slept under their bushes, though photographs were scarce. A friend circulated a convincing picture of her black dog, a (slightly) overweight, shaggy border collie standing in the woods, posing as the bear. The daily paper kept an itinerary of his sightings that offered a sort of map of his whereabouts. Once he crossed the canal, by land or sea, no one was sure; he seemed to be making a beeline for Provincetown. It was a testament to all that no one interfered with his progress and that he didn't get hurt. We cheered him on, hoping he reached his destination unharmed. Left to his own devices, how far would he walk, we all wondered? Would he stay?

Like everyone else, we wanted to see him, and didn't. We waited early one morning by a roadblock the police had set up next to a marsh. Dunkin' Donuts was handing out free doughnuts to a sizeable crowd. Police lights were flashing, staccato voices cutting in and out of walkie-talkies. But he never showed. He had been seen that very dawn by a custodian crossing the church parking lot nearby. Now, according to our calculations, in order to get to that church parking lot, if he had kept to his previous straight-line trajectory, he would have walked right across our farm. The farm was in the direct line of his path from the previous sighting to this sighting. He had never veered off course before. Perhaps in those darkest hours before dawn he had been sleeping curled up under a blueberry bush; bears are notorious blueberry foragers. Alas, lucky for us, but not for him, there are no blueberries in May. He made it all the way to Provincetown before doubling back and being darted in someone's yard. They at least got a picture. He was re-located off Cape where we heard there were further misadventures. Fondly known as the Cape Cod Bear, we called him Henry, after Thoreau (and Beston) who had ambled across the Cape himself, all the way to Provincetown and back. We are not saying that he did sleep in our blueberry field, or even cross it, only that it's possible, even highly probable. But that's our bear story.

JUNE

In which berries form . . .

The dew is still heavy on the meadow grass, which bends under its gracious weight, and small crystal pearls glisten in the morning light. All traces will be gone long before noon. Such is a June morning in the field, warmer now and green everywhere we look. The trees surround the field with a heavy, green drapery of deep shadow and dappled light, and when the breeze moves through the fluttering leaves we realize its soothing sound is one we had nearly forgotten. The small berries forming on the blueberry plants are the same shade of green as their tender leaves. No bigger than pebbles, they cluster under these leafy, protective wings. Every day they grow stronger. There is uniformity to them now, but soon they will array themselves in sea-glass colors of celadon and rose, touched with pale orchid here and a deeper violet there. Small, five-pointed stars, rose-tipped, take shape from their centers. This is the calyx, where the blossom dropped off after pollination. We think of it as a birthmark, a flower-mark, like a watermark on fine paper. We are nearing summer.

The blueberry blossoms, so abundant only weeks ago, are giving way to berries up and down the field. Just brushing against a bush is enough to cast spent flowers to the ground, leaving small scatterings of petals in our wake. As gravity nudges them earthward, their subtle, collected fragrance grows fainter. It is far diminished now by the fragrant, invasive multiflora rose whose scent lingers along roadsides in the warm June air. The dewberry, another white-flowered rose, is an outnumbered native, but it still clusters among the wild grasses in our field with its petite companions, toadflax, and cinquefoil. Hawkweed and red sorrell crowd along the netting posts and run wild in the meadow, turning on its long stems to catch the sun. Only the stands of stately wild blue lupine hold themselves apart on an incline at the head of the wide meadow. We know where to look for them, as we do each plant in its season. Crickets thrum in the long grass. Dragonflies dart past. It's June. All of life is astir.

All we need this month is plenty of warm sunshine and enough water so the plants can draw up nutrients from the soil while drawing down the sun's energy. Their roots are strong and well tested, but their leaves are tender and new. This is a critical period for the plants as they synthesize and transform their energy sources with their own botanical alchemy. We cannot call forth the sun, nor do we know any rain dances, but we can and do supplement rainfall with irrigation, twice per week at this time of year. But even if all the conditions are perfect, with soil, sun, and water, the plants would simply be green shrubs without flowers, and therefore without berries, if it were not for bees. "It's all I have to bring today—this, and my heart beside," to quote Emily Dickinson, an avid gardener, more than a century and a half ago: "This, and my heart, and all the fields, and all the meadows wide. Be sure you count, should I forget, someone the sum could tell, this, and my heart, and all the bees which in the clover dwell ("26")."

The relationship between bees and flowers is a symbiotic one of codependent mutuality. We are the greatest beneficiaries of their relationship. After blueberry pollination is complete early this month and the delicate flowers' work is done, they brown and wither and fall to the ground, leaving in their place the beginning

Wild blue lupine at head of wildflower meadow.
BELOW Bumblebee gathers nectar. | Lupine.

of a small green blueberry. Our bees, both local and imported, are then free to pursue new sources of pollen and nectar. And there are many now. As the berries begin to ripen, the nets, covering two entire acres, will go up with the help of able-bodied friends. This is always a beautiful sight and a harbinger of things to come. We time the netting to the shade of the berries. When berries begin to turn blue their waxy surface reflects light, on the ultraviolet spectrum, that can be detected by birds. In the same way the flowers' ultraviolet light attracted pollinators, the berries' ultraviolet light will attract agents of seed dispersal—birds, animals, humans. At this point they need to be protected from birds especially, and for this we must use netting.

We are rather proud of our nets. They consist of yards and yards of fine-mesh plastic, which we stitched together with plastic wire into panels. We have a separate set of nets for the early ripening section and the later ripening section. Each set is 250 feet long and 176 feet wide. It took a bit of experimentation to get the right size netting for our needs. We had to get the math and science right, too. The netting is wrapped in plastic and stored on the netting posts until they are deployed each summer. Once unwrapped, we use ropes, run through metal grommets attached overhead, to pull them across the rows into one immense net. It takes four people, two with pitchforks, to pull the nets along, and one at either side, tugging the bundle of net toward the center. Once we have pulled both sides of the netting to the middle, they are attached along a center seam. The plastic wires are then tied to the long rows of posts that stand along the sides of the field to firmly secure the nets. The posts stand nine feet tall with an additional six feet underground on some, three feet on others. There are 122 posts in all. The netting that drapes over the sides is fastened to the ground with long pieces of plastic tubing and metal stakes pounded into the ground with a mallet. Two doorways diagonally across from each other have a pulley system for opening and closing; once closed, they are secured with small plastic twists. The nets are up for about eight weeks, from full ripening to end of harvest.

We are always careful to watch for birds or other creatures that might become entangled in the netting, but each year there is often a casualty. A crow once became entangled and had to be ferried, carefully, to a local wildlife rehabilitation center. We have two on the Cape. They serve a noble cause, in our opinion, though it is not one shared by purists, perhaps. We were able to return the crow to our field within weeks and can't help but feel the crows have been more gracious to us ever since. He could be any of the ones we see, though he can easily recognize us. Sadly, we have lost a sparrow or two to the netting, who in their frenzy to free themselves became more enmeshed. We have lost a dragonfly on occasion, though most, along with butterflies, can be caught with a butterfly net and freed.

Deploying nets from winter wrap. Steve fastens nets while workers hold it in place overhead. Pitchfork in use pulling nets. | Nets catching the afternoon light.

Casualties always break our hearts, but never more so than the box turtle whose leg got so entangled it was rubbed raw. The wildlife center had to call their vet (who was also ours at the time) to amputate two of his four rear toe claws. We checked on him over the several weeks of his rehabilitation, encouraged by his progress and the combined skill of the veterinarian and the wildlife rehabilitators. We awaited his return. However, to our surprise, when he was deemed healthy enough for release they refused to give him back to us. They did not want him released at the farm, fearing for his future safety. We made calls, one to the state natural resources office. It turns out that wildlife centers, which receive their license and some funding from the state, are required to re-release rehabilitated wildlife back to their place of origin whenever feasible.

We knew that Eastern box turtles were endangered in Massachusetts and that they reside in the area where they are born. They will try to return to their home if removed from it. They generally live fifty years, though some live to be one hundred. Our injured turtle had the red eyes of a male, a gorgeous golden orange batik shell, and growth rings on his plastron (belly) that made him about thirty years old. Within twenty-four hours of our phone calls, we were able to retrieve the turtle. We released him at the same spot where we had found him, and he crawled off quite vigorously into the woods. Our old dog, Audrey, a turtle lover and great turtle tracker, picked up his trail a month later and led us to where he was basking contentedly in the sun. We couldn't have been happier.

Several turtles live on our farm, and we cherish their presence. We make a concerted effort to protect them from farm machinery, netting, and predators. They love blueberries. We leave small openings in the netting at ground level for them to get under. One delighted two small children one summer resting under the blueberry bush where their family was picking.

Farm work continues at a more leisurely pace, perhaps even a turtle's pace, this month. The rows are regularly mowed and weeded. Where once our helper, Tsering, crawled about on hands and knees under the bushes, now Bob and his sons Anthony and Tommy, clad in plastic kneepads, disappear among the rows. We count ourselves fortunate to have helpers who actually enjoy weeding, because there is enough of it here to keep an enthusiast busy on a weekly, sometimes daily, basis. Once the blueberry development has reached this stage there is less of a threat to their survival, but we keep an eye out for signs of damage from insects or disease. The weekly visits and lab reports of Larry Dapsis, our entomologist, kept us ahead of any winter moth infestation that might have occurred last month. And though the threat is nearly over, some may still blow in on their gossamer threads from neighbors' yards. As the tiny green caterpillar, similar to an inchworm, eats his way to maturity, little leaf fragments and waste drop to the ground from the tall trees overhead. It sounds like a light rain. It is hard not to notice evidence of the winter moth as these tiny scraps of chewed leaves flutter down through the air like confetti and land on your head. Gypsy moth caterpillars join the feast

Nets over ripening blueberry bushes. | Netting showing rope and ring pulley.

in June, sometimes returning with a vengeance after decades of only mild infestation—2015 and 2016, were two such years. Many majestic oaks have been defoliated and some had to be cut down. Gypsy moth caterpillars do not bother with blueberries unless there is nothing left to eat. They have been known to eat pine needles when their population is high and other food is scarce. If leaves are stripped from a tree or shrub for three consecutive years, any plant will die; leaves are their means of life.

Photosynthesis is a word most are familiar with, meaning the process of synthesizing light. Blueberry plants are no different from any other plant in that they absorb sunlight through their leaves. They need about eight hours of direct sunlight daily during the growing season to "manufacture" their harvest. Leaves act like small solar panels made of sensitive green pigment, chlorophyll molecules, that absorb the sun's energy and transform it into energy for the plants' survival and growth. Sugars are stored deep within the plant stems for the winter. Oxygen is released into the air as a by-product of photosynthesis; their waste product is essential to our existence. Emerson wrote that creation is always a circular power returning into itself. It is said that no energy has ever been lost, or gained, in the universe (or possibly multiverses), since the Big Bang 13.8 billion years ago, known as the Law of Conservation of Energy. All the energy present at the moment of creation is still here now in this field, and photosynthesis and pollination, among so many other wonders of the life force, are common, daily, extraordinary expressions of this redistribution of energy.

Celadon green berries with rose-tipped calyx. | OPPOSITE Blueberry nosegay.

The almanac tells us this month's moon is the Full Strawberry Moon, marking harvest time for strawberries. The summer solstice returns on June 21, the first day of summer, or the day when sunlight reaches its peak of 15.5 hours. On midsummer nights, our summer constellations of Ursa Major (the Great Bear), home to the Big Dipper, and Ursa Minor (the Little Bear), home to the Little Dipper, reign over the night skies. Nearby is the amber glow of summer's brightest star, Acturus (the Bear Guard), sometimes confused with the North Star, Polaris, which is not as bright but has the distinction of being the still point of the heavens, around which all else spins. The Big Dipper was known as the Plough in older agricultural societies, and this shape can be easily discerned once you try to imagine a hand plow pushed by man or with the aid of a horse. Most remarkable of all, the Milky Way, our home galaxy, becomes more and more visible in the late evening sky. Our planet and sun and all these constellations live in one of its spiral arms. Looking up at the immense cloud of stardust stretching overhead, it is hard to imagine how we can see something so clearly outside of ourselves and at the same time comprehend our own existence within it. Our Milky Way is only one in a cluster of forty constellations; there are billions more in the universe, maybe billions of universes. Physicists tell us the multiverse is nothing more than a sea of oscillating vibrations; we do not live in a world of things at all. Is there any wonder why we love summer nights, when we are drawn out to comprehend the incomprehensible? We are made of the same elements as the stars. They are a timeless touchstone connecting us to the past and to the future. For us, everything is real enough. We have something tangible here closer to the ground.

Green berries predominate.

Fireflies. The first few seem to appear around every solstice, though the evenings must be consistently warm for large numbers to appear. Their bioluminescence is used for courtship, the males streaking high in flight and the females flashing low in grass and bush. Adult fireflies live only as long as these few romantic weeks of summer, the last winged stage of their two-year metamorphosis from underground beetles. It may very well be that there are fewer than we saw in our childhoods. The common firefly in the Northeast, whose light is yellow-green, is the *Photinus pyralis*, though there are 2,000 species worldwide. Their populations are declining worldwide due to loss of habitat, light pollution, firefly harvesting, pesticides, and climate change.

Red clover. | Coreopsis. | BELOW Rabbit's foot clover

The farm is a self-contained world within a certain, unchanging circumference. It is a framed, finite world but, like Blake's infinity in a palm, it holds the eternal. Sometimes when you enter the field from the main road you can see it more clearly. You do not see an unlimited horizon, nor do you see the staggered planes of a rugged wilderness. The limitless horizon of the shore and night sky satisfy a different kind of need. We have made a place that feels and looks like a sanctuary. It supports many diverse forms of life. They have joined us here. It is theirs as well as ours.

On our early evening walk we notice, in the hollow behind the barn, a bed of sensitive fern with its large scalloped fronds. They have multiplied over these many years. There is sweet fern here, too. A friend from the local Wampanoag tribe tells us that rubbing its oils into our skin will act as a natural mosquito repellent. Its scent is sage sweet. The dogs are led on through the woods to one of the farm's open meadows where the grass is soft and thick with clover: tiny round globes of yellow hop clover, bright raspberry globes of red clover, and fuzzy, paw-shaped globes of rabbit's foot. White clover, everywhere present, seems so ordinary in comparison, but this is the only one that has the lucky four-leafed variety. It is an aberration, which makes it rare and therefore lucky to find. DNA is passed on in plants, so there can be hot spots; in other words, if there is one there is probably another one. This has proven true at our farm. Chances are one in a thousand for finding a four-leafed clover, and one in a million for finding one with five or more leaves. We have found so many four, five, and six-leafed clovers here we have lost count of them. Some of them we press; most remain hidden in plain sight.

We look out at this moving tableau, the sway of un-mowed grass, pristine and pastoral, sparked with flowers and clover. Flocks of bluebirds pass overhead in the company of sparrows and finches. A marsh hawk follows lazily behind and they scatter upward in four directions. Soon this small June meadow will be mowed into a parking area for our summer guests, and so we enjoy it while we can. The bluebirds chirp and swing past on their still-startling blue wings, and another day is done.

JULY

In which the blueberries ripen . . .

We make a running start at July. This month we go from a farm life of solitary leanings to one of bustling crowds, from a single tractor moving leisurely down our dirt road, trailing clouds of powdery dust, to one hundred cars moving single file in a billowing dust storm. Our private place is public once more. There may be a little trepidation at first—is everything in place, will the sun shine on our opening day, will the berries be sweet, will there be enough, will the people come? Somehow everything seems to work out just fine once the gate is swung open. But before that happens, there is work to be done.

We have arrived at the still point of our year—the first two weeks of July—when we await the blueberries' final turn to full color and plump sweetness. Everything has led to this moment. Everything revolves around it. This time is not one of idle waiting, however. Gone are the days of inviting our souls to loaf about, as Whitman entreats us to do: "I loafe and invite my soul, I lean and loafe at my ease observing a spear of summer grass" (*Leaves of Grass*).

If we are methodical, all the details will be attended to. We start with the nets that were put up at the end of June; these must all be tacked to the ground along their lengths. This safeguards them from marauding creatures—deer, fox, coyote, skunk, rabbit, turtle, and all manner of birds, including gangs of turkey, who would rather saunter along than fly. What few birds do make their way in, and they have their mysterious ways, can usually be herded out an opening.

Every July we discover holes in the nets that we didn't get to at the end of last season. They must be mended. It seems that there are always new holes, great gaping holes made by vaulting deer, small bird-sized tears high aloft, and the ground-level tears of squirrels. Gray squirrels are notorious net renders, throwing themselves bodily at the nets, bouncing along its edges until they find a weak spot and rip through with tooth and claw. Some lower openings are left un-mended for these escapades and for our slow moving resident turtles that might otherwise, and occasionally do, dig under the netting rather than fight their way through it. We sew patches of netting on with oiled twine threaded through a fisherman's spindle. It is as tedious as it sounds. When a breeze picks up in late afternoon, the sound of it moving through the netting is a low, whispering whistle of open spaces and sweeping prairie grass. Fine music to sew by, whether atop a ladder or seated on a low bench, stopping now and then to re-thread the spindle or pull the wagon to another hole.

Ripening in the rows.

There is no clear demarcation to mark the end of one month and the beginning of another; calendars are for humans, not for nature where edges blur. But still, there is a point when each month becomes itself, with its own distinctive characteristics that are apparent, generally at mid-month. No matter what the weather, how harsh the winter, how cool and dry the spring, by mid-July the berries are ripe. There they are, in all their crayon-blue finery. Naturally, not all ripen at once on any bush, which would be an unproductive survival strategy for them, as well as us. In our original planning for the field, we took ripening timetables into account in our choice of varieties and planted five different blueberries chosen for hardiness, consistent sweetness, and variable ripening. At our farm we have Dukes, Spartans, Coville, Blue Gold, and Blue Crop. As a rule, in any cluster of each variety there will be a few deep blue berries, larger than the others, which were the first ones to set, and several smaller berries, in various stages of ripening, in shades of paler blue, flecked with lavender and pink. The very smallest berries, the last to form, are a decidedly unripe green. Even if the berries are blue and look ready to eat, it actually takes one to two more weeks for them to sweeten under the sun's warming rays.

What was set in motion, with the setting of buds late last fall, all that was hidden, protected and worked toward, comes to fruition now. Fruition as defined by the Merriam-Webster dictionary is "the state of being real or complete" and "a state of bearing fruit; or realization." Such a long incubation of nine months! The metamorphosis is slow and measured, with strength building upon strength, from bud to flower to fruit. Summer haunts the air, the berries blue under a sky universally blue. Heat rises in wavering mirage. The nearer trees are dark and green and cool. But how we long to throw ourselves down in the long summer grass as the dogs do, as we once did when we were that small and free. We celebrate, in our own way, the wild grasses in all their variety of shape, size, and color, in their beautiful diversity. They provide habitat and food for our resident wildlife. We let them grow wild and mow only some areas when necessary, and it is necessary now, for parking. Once mowed there is a symmetry and neatness, we admit, to the inviting flatness of the temporary parking areas we must create every summer. We always keep our large wildflower meadow to the south, and two wildflower patches bordering the netted blueberries on the east and west, un-mowed. We carve into them a bit to create enough parking, with minimal loss of habitat. Only the small, grassy meadow by the farm table is sacrificed. As opening day approaches, we hammer in stakes, measure off car lengths with ropes, and tie on red flags. Hand-painted parking signs and small white arrows will mark their boundaries for now.

OPPOSITE Blueberries almost ready to eat. | Ripe berries among still ripening.

Perhaps the most important task in early July, certainly the most visible, is putting up our sign. This is something everyone waits for—it is a sign, in every sense, that our opening is not far off. The sign was hand-carved for us many years ago by an old family friend, and it's heavy enough to require two able-bodied people and a tractor with front-loader attached, to defy gravity long enough to hang. We have a set of small (some say too small) individual hand-painted signs for the days of the week—MON, for example, or SAT, which we hang from hooks on the bottom. Passersby also look out for these. There are two special-occasion signs as well—CLOSED FOR RIPENING and CLOSED FOR THE SEASON, neither of which is very popular. At one time in the history of the farm, and not too long ago, this was the only way patrons knew we were open. They drove by our sign. Now they can access our schedule anytime by cell phone or computer.

Down in the field by our newly mowed parking area is a space we leave open for our farm tables. A few years back we built an open structure there that we cover with a canopy. It works remarkably well to shade our patrons when they come out of the field, and us, while we box and bag their berries. The heat of the field and the coolness of the shade create a temperature inversion that makes its own breeze. This trick has been used for centuries by prairie and desert dwellers. Once the canopy is in place, we bring out the tables and chairs and assorted paraphernalia that will stay out there for several weeks, stored in plastic tubs. We haul out the old green wagon, the same one we used for pruning, but this time it holds all the repaired and hand-scrubbed aluminum trays we provide for berry picking. These are re-cleaned after each use and, when the day is done, stacked in the wagon again and covered with a tarp.

A few days before opening, we call all our helpers to make sure they can be there. These days we need about ten. All but two were customers once. They came and never left. Some have come and gone and others have taken their place. They love the farm as if it were their own, and frankly we couldn't do it without them anymore. We pay them with blueberries and our abundant gratitude, a barter system that works both ways.

ABOVE Ray oversees the pickers.
OPPOSITE Farm sign goes up at road. | David stacks carrying trays with pint containers.
BELOW Pickers assemble.

One of our helpers, Bob, is a gentle soul with a deep sense of the absurd. We love this about him, and that he does not mind weeding. He is a great, very funny storyteller. Of all the farm helpers, he has the most enviable encounters with wildlife, from giant snakes blocking the footpath to packs of running coyotes. Our favorite story, though, is the one about the mulch pile. As Bob tells it, he was tooling along in the small tractor about one o'clock on a July afternoon two days before we opened. It was extremely hot and he had the air conditioning on. He had been out in the field since seven and had just come from erecting the farm sign out on the road with Dave. Fairly swooning in the humidity, he was calling it a day when something caught his eye as he approached the mulch pile, not six feet away. What he saw made him bring the tractor to a dead halt and cut the engine. Standing on top of the pile was a majestic stag with enormous antlers. The buck was just standing there in broad daylight. Bob could not believe his eyes. He had never seen anything like this so close. They looked at each other for a while. Time stood still for Bob. Then the deer leapt off the pile, twitching his white tail, and bounded in a zigzag pattern into the woods.

Bob turned the engine over and continued back to the barn, where he told Dave, "You aren't going to believe what I just saw." His tale was told, more than once, with awe rather than his typical humor, but the mulch pile added a flourish of the absurd. This highest point on the farm has since been dubbed Hokum Peak, in no disrespect for Black Elk Peak (formerly called Harney Peak), the sacred place of the Lakota Sioux in the heart of the Black Hills of South Dakota (where one goes to cry for a vision and hopefully be visited by a spirit animal). To add a still finer point to his story, in Celtic mythology the stag is the King of the Forest, protector of all creatures and caretaker of Mother Earth. He could not have arrived at a more propitious time if we had invoked him, which of course we had.

The day before opening for the season we go down our checklist again. We get cash for the cash box and ice for the thermal water jug we put out. On a whim some years ago we designed and had made for us a small blueberry stamp. The stamp is a temporary tattoo for children. We make sure it's out by the table or there would be a lot of unhappy faces. We check the ink pad, twice. Fresh flowers for the table are picked from our garden, sometimes garnished with a friend's flowers. Paper bags and carrying cartons and box upon box of pint containers, called punnets, are hauled out to the field. We make sure a month ahead that we have plenty of these from the distributor. We pick up more business cards and "Buy Fresh/Buy Local" farm guides, magnets, and bumper stickers, too. The weekday sign is checked out front, we confirm the opening on our phone message, and most importantly now, on our website. The rows are walked and any broken branches snipped off. We mow again. On the eve of opening day we walk the field one more time, drawing a protective circle around it. We talk to the berries. We tell them this is their time to shine. We thank them, we encourage them, we tell them how beautiful they look and are. And we say our goodbyes.

It might seem like a lot of preparation to open a blueberry farm, but you'd understand if you came down on opening day when the gates open and the blueberry lovers flood into the farm. Patient no more!

Our all-volunteer staff is on hand by 7:30 a.m. to prep and stack the trays with punnets. We catch up on each other's news. We all feel a bit nervous about what to expect. Aprons and vests are distributed. When our helper Ray heads off to the gate under his straw hat and gaily-striped oversized golf umbrella, silence descends. Everyone looks at their watches. Some dash into place. Some stand glued to the spot as if awaiting a deluge. There is nervous laughter. Then the cars roll past. One after the other after the other with no end in sight. Some have been in line out on the road for an hour. The first cars in are, without exception, the same patrons every year. Most of these early birds are old timers with us now. All trepidation flies away. This is not a marauding war party; these are all our friends. We know most of these people smiling and waving as they steer into their parking spaces and descend on the table. There they form a line, single or double file, with their trays and baskets and buckets at the ready. Steve gives a welcoming speech and then they're off. Out into the rows.

Polka-dot picker.

ABOVE Blueberry tattoo gang.
A pair of low-pickers. | Neddy admires a plucked berry. | Farm table ready for patrons.

Once the first wave has passed and everyone is out in the field, there is a lull for about twenty minutes, and then the second rush of traffic arrives. Parking and traffic control has become an art. Wooden signs and arrows direct parkers, and so do the flagging arms of traffic controllers, but with all the pedestrians, small children, and often dogs, there is a lot to keep an eye on. After that, we see a steady flow of cars every few minutes or so. On busy days, overflow cars park around the checkout table and along the meadow, the closest stops reserved for those with handicap signs on their dashboards. Wheelchairs and strollers are easily pushed out into the rows, and we place canvas camp chairs out for those who tire easily or just want to watch. Generally there are four or five helpers in the field with the pickers. We hope the picking moves from row to row in a bit of a uniform way so we can keep track of what rows have been picked clean of ripe berries and which ones haven't. That is the best practice for us, and for the berries and the pickers, too. Berries should be easy to find and pick, and they are.

Many of our visitors have been coming to the farm for years and bring their own baskets and picking tins. Some are passed down from one generation to another. We see an unusual array of inventive ideas, from coffee cans on strings to plastic milk cartons with their tops cut off and holes punched in the sides for neck straps, and Tupperware in all sizes. One gentleman brings the same plastic pitcher every summer that he holds under the branch he is picking from until the pitcher is full. He is there every day we're open, and if he's not, we wonder where he is. In the (good) old days, most pickers brought their own containers, but as the number of patrons has grown over the years, so has our provisioning; many have never picked blueberries before, so it is a new tradition for them and they need our guidance. There is a steady stream of outgoing and incoming traffic, new arrivals gathering to pick up trays, crowds lining up to pay at the tables, chatting, sipping water, and getting tattoos—an endlessly (three-hour) revolving circle. Each of our helpers seems to anticipate the needs of everyone around them. It may be hectic, but it is no more stressful than a blueberry social.

Once the gate is opened for the season, we usually open three days the first and second weeks, then only two days for the last four weeks, and only in the mornings. It gets hot out in the field by noon, and we couldn't ask our helpers to lend us more than three hours of their spare time. As always, openings are weather dependent. Sometimes we close for an entire week of ripening mid-season. We gauge this on a day-to-day basis. There is a fine balance to maintain; there have to be enough berries for all the patrons, but they have to be plump and sweet enough to harvest. There will always be some loss—some berries will over-ripen, be eaten by birds, or fall to the ground. Each bush yields about six pints of blueberries. At our current price of $4 per pint, one might think there is a high profit margin in blueberry farming, but expenses run high and we are happy if we break even every year.

We are modestly famous now—on a local level that is, though we do have customers from all around the world who vacation on the Cape. Articles have been written about us in local newspapers and magazines, interviews broadcast on the local NPR station and local cable. New customers tell us they heard about our farm in the oddest places—sitting at a bar, working out at a gym, in the ladies' room of a restaurant, in the aisle of a grocery store, on a whale watch cruise. Some young people and families have come our way through a Google search that turned up our website. The Internet has added to the word-of-mouth business that sustained our earlier years. We ask ourselves every summer if we have reached our limit, the carrying capacity of our farm. But we never answer that question. There are only so many berries; somehow everyone seems to get what they need. We know so many people now because of the farm and we know how much it means to them that it is part of their lives. It wouldn't be summer without it. Sometimes we get gifts. Our helper, Mary Jane, made us a picture-perfect blueberry tart. One patron who vacations here every July with his family (now our close friends), makes us a delicious blueberry pie from his grandmother's recipe every summer. A cranberry-growing friend makes us the most incredible jam combining our berries with her own. Patrons often share their favorite recipes, which we post on our website. Picking blueberries on a summer morning simply makes people happy; even the berries seem to glow. July is that kind of month that seems to go by in a day. The blueberries have made us into social butterflies, flitting from field to table from the time we open the gate until dusk.

There is no doubt that blueberries are good for your health. All studies point to this fact, and new research in neuroscience shows that contact with nature changes brain chemistry, releasing those positive, euphoria-producing hormones called endorphins. That is all well and good, but there are other intangibles to consider. Imagine, for a moment, you are standing in a field on a summer morning with family and friends. You are picking sun-warmed blueberries. You pop one into your mouth. It's sweet and moist and delicious. Crickets keep a slow, steady beat in the swaying grass. A bluebird flits overhead. You take a deep breath, in and out. There is no other place in the world you'd rather be. That could be called contentment, or, for us, blueberry bliss.

OPPOSITE David, Matthew, Ben, and Andrew visit every July. *Photo courtesy of David German*
Buy fresh, buy local. | A bucket of berries. | A bowl of berries.

CAPE COD
BUY FRESH
BUY LOCAL

Milkweed blossoming.

In our wildflower meadow, these hot summer days are given over to the sun-loving flowers—the yellow-centered daisy, black-eyed Susan, coreopsis, and tall, yellow-crowned mullein. We hear the cries of the soaring red-tailed hawk and the squawking response of a blue-black crow as they continue a dispute neither will ever win. "The bluebird carries the sky on its back," Thoreau observed (in a book of essays of the same title), and flocks seem to materialize from the sky itself to flash from post to ground and back. A patron once asked, after seeing so many in the field, if there was any connection between bluebirds and blueberries. Not as far as we know through science, though the evidence is clear to any eye here. What they love is the open space surrounded by protective trees with plenty of resources to choose from. Their nests are in tree cavities ten feet above the ground, and they can have two or three broods each season. Their eggs are very small, a half-inch in diameter, and of the palest blue color. A bluebird can live for ten years, though most have far shorter lives. Once threatened by competitive introduced species, such as sparrows and starlings, in the early twentieth century, there has been a healthy population increase since the 1960s. They eat insects, grass seed, and blueberries when they can get them, which provide essential fluids and nutrients.

The field of summer stars seems closer to the earth in July, the constellations so near we can almost reach up and touch them. If you look at a blueberry you will notice that each one is marked with a five-pointed star where its flower once bloomed. The Algonquin-speaking tribes of the area called blueberries star berries. When the Full Buck Moon shines the constellations retreat to their vast distances again, as on an outgoing tide. We hear the tremolo of the screech owl hidden in the woods and the sonorous pitch of the great horned owl that seems to echo in the vast chamber of its chest into our own. Bats swoop past too close and fireflies not close enough. There are rustlings among the trees that could be any shy night creature. We remark to ourselves that we try to do too much in summer. The days are at their longest, but they move too quickly. They are already turning us toward August.

Butterfly weed. | Milkweed with long horn beetle (red milkweed borer).
BELOW Milkweed blossoms.

AUGUST

In which the harvest is complete . . .

We can feel the year turning now, as if the sands of summer are rushing through a narrow hourglass neck. Canada geese gather in the ponds, too early it seems, with goslings on full wing. Fall's goldenrod appears in the field, too soon. We anticipate the return of our fall grasses every August, as if in consolation for what we are about to lose. "I believe a leaf of grass is no less than the journey-work of the stars," Walt Whitman wrote in *Leaves of Grass*. Islands of grass are on the rise on either side of the netting as shadows lengthen. Bumblebees nap amid the folded petals of black-eyed Susan and the broad leaves of milkweed. We are six weeks past the summer solstice as the month begins, and we can feel a softening of the light. Twilight's approach is earlier every evening, and by mid-month two hours of sunlight will already be lost. August 1, was traditionally known as Lammas Day, the day of first harvest. Our harvest is halfway done—or is it only half begun?

Blueberry morning. | OPPOSITE August berries.

We are as busy as we were last month. We are generally open for three weeks in July, and we have two or three weeks left in the season. By now most of our patrons have been here more than once, some every time the gates are open. Although the atmosphere is still festive, some patrons have turned a more serious eye toward the harvest. Gathering enough berries to feed themselves and their visitors seems secondary to putting away enough in the freezer to get them through the winter. It is as if, for some, the first fruits were squandered in pleasure and these last are to be hoarded. Some tell us, half in jest, that they kept our farm a secret for many years, but it is a secret no longer. Our visitors have many different styles of picking; some crawl on the ground, some concentrate on a single bush, some graze from one to another. Our helpers are always at hand to show them where the best berries can be found. We encourage tasting among the different varieties, and we try to teach children, with parents' encouragement, the difference between blue and green berries. Dogs are always welcome if they are friendly. Patrons alert us to things—pickers who are hogging bushes, which we don't allow; bird nests with eggs or hatchlings hidden among the blueberry branches, a box turtle beneath; a dragonfly caught in the net. One year a baby cottontail was discovered, prostrate on the ground, breathing heavily in the ninety-degree heat. The little rabbit had strayed too far from burrow and family. We placed him in a basket in a nest of grass and rushed him off to care, stroking the top of his velvety head. It is quite often children who notice these things. They line up at the checkout table, hands extended, to receive their blueberry tattoo when picking is done. We are never too busy to neglect our duties with the little wooden stamp and inkpad and are occasionally reminded if we have overlooked them too long. The temporary tattoo looks best on a child's small hand, though many parents request one for themselves, stamped carefully amid freckles, hairs, and ropy veins.

OPPOSITE Picking continues.
ABOVE Regulars Sylvie, Guy, and grandson Taz. | Regulars Peter and Ellie.
Annie, a regular, makes jam with our blueberries and her cranberries.

What exactly is a blueberry? It is a flowering perennial plant or shrub of the genus *Vaccinium*. The Northern highbush blueberry is *Vaccinium corybosum*. There are twenty-six species of *Vaccinium*, including cranberries. All *Vaccinium* fall under the family known as Ericaceae, which includes rhododendron, azalea, heath, and, surprisingly, the eerie Indian pipe, or ghost plant (Emily Dickinson's favorite) just emerging now in the August woods. Berries vary in size from 5–16 mm, or 0.20–0.63 inches. The largest we have seen on our farm measured at least one inch in diameter. An excited young boy once showed us a berry that looked considerably larger than that. His parents later told us he insisted on mailing it to his grandparents. Many patrons notice a whitish, waxy coating on some of the ripest berries. This is known as bloom. It is a natural epicuticular (surface) wax that protects the berries from moisture and insects. It is a sign of freshness and does rub or wash off. Interestingly, it is the bloom's ultraviolet glow that attracts birds, which, like insects, can see on this spectrum, which is beyond our abilities—this is the science of blueberries. But we experience their attractive force, too, even though we do not see in that broader spectrum of light. Their round shape and beautiful blue color attracts us, and the way they roll in the palm and taste on the tongue. People have a hard time stopping when they start picking or eating them. Looking for them is like a treasure hunt. They have some kind of magic, luring us on to pick just a few more. They cast a spell on us all.

Hidden blueberries.
OPPOSITE Blueberry field in late afternoon.
Late blues. | Blueberry colander.

A blueberry plant, you may recall, reaches maturity at eight years; ours were planted in 1997, which makes them more than twenty years old. A healthy, well-established blueberry can thrive for decades, barring disease and insect invasions. We have been fortunate in that regard. We work hard to give them all the care they need to do well. We are not planning on planting any more bushes, though recent business could sustain double the number that would require twice the effort, with more workers and helpers. So we are in the enviable, or unenviable, position of being too successful. There is no going back to the old days, but how do we move forward into the future? Do we simply maintain our fields as they are with their attendant rush of customers? Blueberries will not be losing their popularity anytime soon. Perhaps in the future there will be others to plant more berries. Perhaps other farms will catch up. We enjoy being on the more profitable side of the production most of the time, after all these years, but we are fully aware it can end anytime, too.

In the wildflower meadow, spotted knapweed is one of our favorite wildflowers that return this month. Some people may call them weeds, but we think that insults their natural beauty. They can make even the dreariest roadside ditch come alive with their spiky purple flowers, brightening the grasses as they brown in the summer heat. Bees love them. Milkweed plants grow tall and sturdy and verdantly green all across the wildflower meadow, holding their own against the invading mugwort, which has nearly taken over. Mugwort is a bother to us, yet the plant has its enthusiasts, long valued by herbalists for its many medicinal uses. There is one inescapable similarity between mugwort and milkweed. Mugwort has evolved a chemical it emits to prevent many other plants from thriving in close proximity, which is why, once established, they crowd other plants out. Milkweed has evolved a milky chemical of its own, a toxin that, once ingested by insects such as the monarch larvae, protects it from being eaten by predators.

The milkweed multiplies alongside the mugwort. It raises its proud, tasseled head above the fray, its globe of flowers a burgundy and cream enticement. Late last month, the air was filled with its heady aroma, a coconut honeysuckle scent, not overly sweet but rather pleasant. Theirs is an elegantly rich perfume, inviting an unusual diversity of pollinators. Once pollinated, the clutch of fertilized flowers will go as limp as old brown rags, their perfume no longer wafting in the warm air, and the slender horn of their seed pod will emerge. On higher ground, the wild lupine has gone to seed, with fuzzy, dangling pods that will dry in the hot summer sun and curl back as the weeks pass, releasing seeds for next year's growth. As one plant finishes its cycle, another's has already begun. Taking its colorful place is the brilliant orange flame of the butterfly weed, bursting into bloom as if from nowhere, and the emerging monarchs are soon to follow.

OPPOSITE Milkweed pod.
Spotted knapweed with bumblebee. | Milkweed tussock moth caterpillar eating milkweed.

As the milkweed pods began to proliferate, we discovered a little critter eating the plants, much to our distress, which led us off to another entomology lesson. We learned its name was the milkweed tussock, or tiger moth, whose larvae feeds on the milkweed plant, just as monarch larvae do, though we have yet to see them. During its short, happy life as a caterpillar, all the tussock does is eat, often working a plant with its siblings, their feathered headdresses bobbing as they inch up and down the stalks. Here one day and gone the next, these tiny, fast-moving Halloween creatures leave only skeletal remains in their wake. This fussy caterpillar, however, feeds only on the older milkweed shoots, leaving new growth, generously, for its celebrated cousin, the monarch. Both feed solely on the milkweed plant, ingesting its mildly poisonous, milky sap, which protects them (somewhat) from predators for their entire lives. In their larval stage, both caterpillars are orange and black striped, nature's warning colors that alert predators to the presence of toxins. Only the monarch will retain this bold coloration; the conciliatory moth is rather drab in adulthood. Moths and butterflies, *Lepidoptera*, make up the fourth largest order of insects. Loss of habitat and overuse of pesticides has reduced milkweed populations nationwide, affecting the monarch butterfly and no doubt the milkweed moth populations. Efforts are being made to gather seeds from threatened native plants and to encourage their planting and protection. Our population of milkweed plants remains strong so far, though we have noted a decline in the number of monarchs every year.

Clouds lift off the horizon as evening settles, forming a parallel horizon. Beneath is a band of pale orange pastel etched with the black outline of tall trees. The light seems held aloft momentarily, there and in the billowing yards of netting on their steady posts. They shimmer with a silvery glow, both solid and airy, with a sense of something captured and stilled. Bands of lavender are now visible in the lowest gray clouds beyond, and the uppermost are rimmed with white light, backlit, as the sun descends on another summer day. We stay to watch the last full moon of summer rising as the summer sun sets, the Sturgeon Moon, known to some as the Blueberry Moon. It is a Chinese lantern rising from behind the tree line, inclining its face over the dark fields, deep in cricket song. We can almost watch it move. Its candlelight glow is as languid as the sultry liquid air that clings to our skin. Later it will be a bright white orb, distant and cold, but now its sad orange face holds onto the last of the sun's warm rays. There is a face in the moon so clear and apparent, almost a presence; it seems lit from within, but we know that is the moon's illusion, her trickery. She is a mirror only. Does her face reflect something seen below, some sad sight—the approach of autumn toward our small blue field, perhaps?

Every year we hope for the best of seasons, but there are variables we have little control over—the weather for one, insects for another. As we look back through our farm journals, we will remember the summers of 2014 and 2015, as especially bountiful, whereas 2005 was our earliest and shortest season on record. Damage from winter moths was extensive, with ninety percent of the crop lost.

Hawk talon entanglement. | Steve repairs net.
BELOW Trusting hawk re-released ten days later.

Early, excessive, and prolonged heat and humidity pushed the ripening too quickly to stagger over several weeks and left us vulnerable to infestation and disease. No year is quite like another. In our journal we noted that during the 2013 season we experienced the long-anticipated and much-dreaded reappearance of the spotted wing drosophila. This invasive fruit fly, native to Asia, first surfaced in California in 2008, and rapidly made its way across the country, arriving in New England in 2011. It made its first appearance on the Cape, and at our farm on July 23, 2012, spoiling about thirty percent of our final crop that first year. Their effect has not been as strong since, because we keep a watchful eye on them, with the assistance of our entomologist, and spray when necessary. The drought of 2016, seemed to have kept their numbers down. Unlike the nuisance fruit flies that we are familiar with, hovering on over-ripe fruit, this invasive attacks a fruit from first color until it is ready for harvest. The flies multiply rapidly. Other fruits besides blueberries are also at risk, such as raspberries, grapes, peaches, apples, and even tomatoes. There is no natural method of control known at present, but nationwide entomologists are scrambling to find solutions; our farm samples are part of the statistics they work from. Research into effective methods of control and eradication will hopefully follow as knowledge builds, but for the time being we employ the best practices of integrated pest management (IPM) pesticide application, our only source of protection. As Larry says, "You can't feed the world with organic farming."

It's a wrap: end-of-season net wrapping atop special platform.

The support of our county entomologist has become increasingly important to us. His visits make us more fully aware that we are part of something larger than ourselves; we are part of a nationwide agricultural community with a long history and an unknowable future. We always learn something from Larry, and his enthusiasm and dedication is boundless. Every farmer should be so lucky. Various trapping devices of his are now commonplace out in the field—small red plastic cups and green and yellow tented cylinders. They dangle from metal stakes planted inconspicuously, or not, among the bushes. He installs them on many of the 476 farms on the Cape and makes his rounds regularly. We have three. The word trap is a misnomer, as trapping the invading winter moth and the spotted wing drosophila would be impossible. The traps simply provide early evidence of their presence, giving us a timely advantage in getting rid of them. Lurking on the future's horizon is another fruit-loving invasive from Asia that arrived two decades ago, the brown marmorated stink bug. There has been a dramatic increase in its range and destructive behavior in recent years. Live samples were trapped on our farm in late August 2016.

As climate change becomes more widespread, we can expect that new insects that are harmful to local crops will become more of a problem, among myriad other challenges. Invasive species can also move from state to state and country to country through normal channels of travel, transport, and importation. The loss of bees, butterflies, and even fireflies is more and more apparent to our eyes. Some science points to a direct correlation between the rise of these insects and the extinction of so many bird species. Because we have worked so closely with entomologists over the years, and never more so than recently, we are becoming increasingly fascinated with and concerned about this world within the world of the farm. Indeed, if predictions prove true and there is an increase in average yearly temperatures, the highbush blueberries we grow now will eventually have to be replaced with southern lowbush varieties, which do not require an overwintering chill time, if we don't lose them first to insects.

Everyone can feel the last day of the season coming. By mid-August, fall is stirring in the air. We are on the second sweep of the field, in the fifteen rows of later-ripening blues. Picking has slowed a bit. There are patrons who are out in the field nearly every time we are open. Some patrons have already said their goodbyes. Indeed, some move away and we don't think we'll see them again. Others are just passing through this once. We occasionally have "private picks" for those who are only in town on off-days, those whose blueberry needs are few—a family from Chicago, another from DC, a friend from far-away Falmouth. Regardless, attachments have formed over the years. Many are good friends now, many others friendly acquaintances we will run into here and there throughout the winter. We have seen children grow into adults, and we have accepted the loss of many farm friends who have passed on. We are part of each other's lives. Saying goodbye means another summer is nearly over, as well as another year in our collective lives. The simple truth of it is that we will all miss being out there; we will miss each other and these casual morning get-togethers. The blueberries have brought us together and we will miss them, too. Affections run deep. We might wish for these fields of summer to last all year. We might wish for the good things we have created and the relationships formed to last forever, unchanged. But we know this is not possible. Impermanence is the only real constant in this natural world.

On our last day of the season, we often impose a limit of one or two pints per person, just to be fair. We lose some of our crowds with this edict but the diehards remain, grateful, as we are, that there is enough to go around. Finally, the "Closed for the Season" sign is hung out front, the gate closing with a forlorn creak, and slowly, slowly, we re-pack everything with care, take down the canopy, pull the wagon back to the barn, and another summer is gone. We wait a while before we take the sign down, just so the transition doesn't seem so abrupt. The bare space between the pine trees by the road can look empty without it.

Meanwhile, the approach of autumn becomes more apparent every day. Late summer is hurricane season, and we try to get the nets down as quickly as possible when the farm has closed, though there is considerable mending to do before that can be accomplished. We often need to complete repairs on the net that we didn't get to before we opened. Sometimes there are new rips and tears. Weather permitting, if no storms are in sight, we will try to finish this up. If not, it will go back on the list for early next summer. Until we finish, we leave the net "doorways" open for the birds, even if berries are few and far between. Bob's weeding, sometimes with the help of his sons, will take them into October.

As the summer swirls to a close around us, with all its frenetic rush and noise, we are returned to our peaceful solitude. Our workers weed the rows in the morning on hand and knee and we join them in the lazy afternoons to sew nets. Birds twirl and glide past us on the currents of warm air. Above us, blue-backed swallows materialize. They sweep past with wings outstretched, white and orange breasts flashing just inches from our heads. Tilting the rudder of their long, forked tails, they veer off with a cheerful call. It is a small battalion we watch; even the dogs tilt their heads up and gaze quizzically at the show. Cumulous clouds are banking in the blue sky. Doves roost in the trees, and the sweet music of crickets reminds us that late summer is indeed here.

A few days after we closed one year, and before the nets were taken down, we had a "netting" we will long remember. One of our helpers discovered a red-tailed hawk hanging upside down, its talon wrapped in the high net. Fortunately he was on the inside, though we had no idea the hawks would venture into the netted field. He hadn't been there very long. He was alert but passive enough to allow us to clip the net around him, free his talon, and then wrap him in a blanket for transport. Those nets—"You can't live with them and you can't live without them," the helper wryly observed. Another trip to the Cape Wildlife Center, and two weeks later, re-hydrated, his talon healed, the hawk was re-released in the field at the spot where we'd found him. The nets were long gone. They told us he was a juvenile male. Another hawk had been sitting in the trees calling to him as we cut his talon free, and kept vigil there every day for a week. We do not know if it was his mate or his mother. We hope they found their way back to each other if that was meant to be. It's an odd coincidence that the two birds caught in our nets, rehabilitated and released, were arch enemies, found at opposite ends of the field.

Once the netting repairs are done, the stakes are pulled up from the ground and the nets pulled back across the field in reverse, rolled up, wrapped in plastic sheeting, and tied to post lines for the winter. We had to build a wooden platform to accommodate a team of three men to accomplish this last task. The nets are ungainly now, hard to wrap after weeks of freedom in the wind; they must be wrestled into submission. The platform is lifted up by the tractor's front-end loader. It takes us about ten hours to put the net up and ten hours to take it down, section by section.

Our gaze returns to the spacious openness of the unfettered fields. But when did the goldenrod turn so golden or spread itself so throughout the field? Monarchs flit merrily past, always in a hurry now, no time to linger, some brightly colored, some tattered and pale. The meadow is textured with sun-worn hues of gold, amber, and fawn, and among summer's faded garments peek the pale pink smartweed, lady's thumb, and the last bruised black-eyed Susan. At the head of the wildflower meadow stands spiky thistle on bristled stem, its deep rose flower a medieval standard with its royal hue. We see the hardy chicory, with its heavenly blue petals, amid a scattering of low asters. They will grow anywhere. Stands of Queen Anne's lace reign above them all, leaning southward, the delicate handiwork of their parasols and nests holding the light. Along the post lines, in swatches of red and gold, gather the last of the red sorrell and withering hawkweed, no longer opening and closing with the day.

There is not one decent blueberry left in the field.

What though the radiance which was once so bright
Be now for ever taken from my sight,
Though nothing can bring back the hour
Of splendor in the grass, of glory in the flower;
We will grieve not, rather find
Strength in what remains behind;
In the primal sympathy
Which having been must ever be . . . .

—William Wordsworth, excerpt from "Ode: Intimations of Immortality" from *Recollections of Early Childhood*

SEPTEMBER

In which fall begins . . .

Autumn begins to weave upon its loom in threads of henna and gold, drawing in leaf and grass and flower, the warp and weft of its shuttle flashing. A whole new cloth will emerge, intricate in pattern, knobby and richly colored. At the west end of the field, toadflax leans lazily among the long stems of browning grass, unperturbed and peaceful. We often find ourselves there standing atop the mountain of leaf mulch and looking back across the rows of blueberry bushes. It is the dogs who lure us here, but we have come to appreciate the view as they set about their primordial digging. From this lofty, if unsteady, height, there is no longer a blush of blue anywhere, only an infinitude of the deepest green. From the earliest days of the month to the last, fall's transformation will unfold. We may have wanted to hold onto summer a little longer, but we are breathing a sigh of relief, too, with every passing day. Summer days have rushed past, dropping us from their heights into this slower realm. Now, at least, and a bit sadly, we have the place to ourselves again.

September's is a quiet beauty, a slow fire burning, with hours of warmth still promised, and it is a month we cherish at the farm. The flocks of swallows continue to divert and amaze with their graceful, quicksilver acrobatics. "They came like swallows," Yeats wrote, "and like swallows went . . . half a dozen in formation there, That seemed to whirl upon a compass-point, Found certainty upon the dreaming air" (Coole Park, 1929). We stand mesmerized. They glide past inches above our heads, up and down the rows, then sweep up in wide concentric arcs. Just as suddenly as they appear, they seem to evaporate right before our eyes, one by one, trailing off as if we had only dreamed them. Of course there is a science to their arrivals and departures. They coincide with fall migration and insect hatchings in our field, including the spotted wing drosophila no doubt, a ready source of sustenance for their travels, but still.

This month we work at a more moderate pace as the days shorten and the grass turns its spectral gold. Too soon there is a chill in the evening air and often mist lingering in the field come morning. A late afternoon fog may sweep up from the shore with its great ocean scent and we feel that we are leagues out to sea. September is a tranquil month with a few fall chores to accomplish as we turn from harvest toward winter. The blueberries will do the same. If we did not finishing wrapping the nets in their heavy plastic coats, tying them along the post wires, we must attend to that now. The tractors are often out, ladders lie in the grass, and seed bags filled with annual rye lean against posts, ready to be tilled

ABOVE The field in September.
Rototilling soil to plant grass seed between rows. | Tilled and ready for seed.

into the rows. It is a good month for these tasks with crickets, summer's songsters, still humming happily in the long grass, and dragonflies, miniature helicopters, snapping past.

At our latitude (41 degrees N, 70 degrees W), the peak period for monarch butterfly migration is September 8, through 20, confirmed by our own sightings. The earliest we see this butterfly in the field is July, their arrival timed to the blooming of the brilliant butterfly weed; in fact, we once saw two mating nearby. Sightings are more frequent in mid-August, more so in September. Although monarchs lay their eggs, 400 in all, on the underside of milkweed plants, we have never identified any. Within five days the eggs hatch out and small caterpillars emerge, smooth skinned and banded with yellow, black, and white stripes. After eating and growing for two weeks, they pupate—that is, form a pale green, gold-spotted chrysalis. Two weeks later, they shape-shift from crawling insects into winged ones. No matter how often we try, we have never seen this entire process first hand, only parts of it. The chrysalis is well camouflaged and hidden; even so, their survival rate during this vulnerable transformation, not withstanding the milkweed toxin, is a low five percent.

Newly emerged butterflies will linger awhile, feeding on flower nectar and gaining strength. In September they will embark on their ancestral migratory journey all the way to Mexico, where colonies winter over by the millions. One fragile monarch cannot make the entire round trip, as a bird might. The butterflies born on our farm will lay eggs on their return trip north before dying. They will never return to our farm, but their great-grandchildren may. When their eggs hatch and pupate, somewhere in the southern leg of their flyway, their offspring will continue the parents' odyssey. It is an amazing migratory feat. If we were to see, and somehow recognize, the offspring of this season's monarchs next year, they would be the third or fourth generation returning to the farm. Wherever we see them in the meadow, flitting about in their carefree dance from flower to flower, we follow; little orange kites, with toxin laced armor and tissue paper wings, we marvel at their power.

As summer surrenders to fall, we continue our irrigation efforts on the blueberries' behalf. They are still growing. As long as the leaves are green, until the end of the month at least, photosynthesis is taking place. In the spring, the plants' resources went into developing new shoots and roots; in the summer, fruit development. Now that the berries are gone, the plant will turn its energy to the growth of new shoots and roots once more. Growth will accelerate while temperatures are still favorable. The plants are also setting next spring's fruit buds. If we look closely, we can see tiny buds forming here and there on the branches. It is an unusual sight to see new buds setting in the fall, and yet it reminds us of how intelligent and hard-working the blueberry plant is—it seems to have little to do with our presence at all. This process will take them into December.

Included in our September chores is weed control. We, or more accurately, Bob and his sons Anthony and Tommy, will continue to weed along the rows until first frost. Early in the month we rototill the corridors between the rows into neat furrows for the annual rye grass we will sow. The rototiller and seeder are hauled behind the small tractor, which is the only one that can fit down the rows. Rye, which is a hardy and fast-growing grass, will sprout within two weeks of being spread. It is a cool-season grass, green in spring and fall, and brown in summer. Warm season grasses, on the other hand, such as switch grass and blue stem, are brown in spring and fall and green in summer. It is always an incongruous sight to see a carpet of spring green grass emerging amid all the autumnal color.

Toadstool gills. | Milkweed pods and seeds.
BELOW Indian pipe: a chlorophyll-free, mushroom-like wildflower.

Marking the end of the traditional farming season is September's full moon, the Harvest Moon, sometimes called the Corn Moon. And the twenty-second marks the autumnal equinox, when day and night are again of equal length. It is always hard to believe that three months have passed since the summer solstice, the longest day of our year. We will have lost three and a half hours of daylight by month's end, which makes September equivalent to March. There is no denying

ABOVE Canada geese foraging in field.
Blueberry leaf with early fall color. | Chevron flight of migrating geese.

the fact that the sun is lower in the sky now and its path from sunrise to sunset is narrowing. In the canted light, the sun-weathered grasses take on a golden glow as if lit from within. It is a color with experience behind it, deeper and richer than spring's bright extravagances. This is their time to shine, and they do, the tall switch grass, the drooping foxtail and wild rye, brilliant assemblies of little bluestem, and the solitary timothy. In contrast, the blueberry leaves still hold most of their summer green though darker and drier now with a mahogany burnish on some leaf edges. By the fourth week, leaves will be dappled with orange, rust, red, and bright yellow as they join the grasses in their autumn display.

Over the summer, the wild turkeys have expanded their brood and dash about with indifference, their heads held high. They never seem to worry about changing their behavior for any season. But overhead, almost daily now, we hear the cries of Canada geese gathering loose families together into practiced flocks. Some will stay in the area, those whose migratory patterns we have interfered with, but most will not. It is the migrants who usually visit the farm on their way south. The resident geese never do; perhaps the greenery of the nearby golf course and the glistening waters of the surrounding ponds and marshes are their more familiar haunts. The ospreys' calls overhead will become fewer with every passing day until we realize that we haven't heard them in a while, and they too will have gone their way. As the days warm and cool, the cricket chorus rises and falls. A soothing cadence no longer, beats are skipped and rhythms lost, then regained, as their numbers dwindle. The catbird will leave soon too, but for now his melodies follow us as we walk. He seems a gregarious bird, always interested in our comings and goings, and quick to run through his repertoire, which we admire. Milkweed pods creak open. Bumblebees drift lazily among the asters and stop to slumber among its close petals. Acorns crunch underfoot and pinecones are scattered along the pathways to be secreted away by squirrels, and sometimes by dogs. Turtles sun themselves as long as they still can before disappearing under shallow mounds of earth and leafy piles in the woods. There are little bite marks on the mushrooms that have materialized everywhere we turn. Their scent is rich with the earth they have pushed up through. There are fairy circles where the farm tables stood.

Rites of passage mark beginnings as well as endings, and we will always remember September for two. One year a young couple came to us in search of a "rustic, working farm" for their wedding. A year later they were married here, under an arbor made from driftwood, shells, and flowers. The bride walked down an aisle of blueberries on her father's arm. Just seeing the white spires of the wedding tent against the blue sky out in the field was a revelation—something they imagined that we never had. Our checkout arbor became an island where ribbons of colored cloth fluttered like waves on the sea. White-clothed tables were placed, just so, along the meadow with vases of flowers and candles. A cozy dressing trailer sat nestled at the edge of the wood. An overturned rowboat was an ice chest. Fish-shaped glass lanterns lined the dirt road, and tiny white lights sparkled from the netting posts. Blueberry cordials were served.

Under the high tent hung strings of paper flowers and lanterns, strings of white light. Guest tables were set with candles, flowers, feathers, and place cards; another table was set for exquisite cakes, another for gifts. There was a dance floor, a stage. We couldn't believe our eyes. The whole effect was both elegant and rustic, transforming our farm into a fairy-tale landscape for a very real, significant, and sacred ceremony. We were deeply moved. These young people, strangers to us, had become the dearest of friends. Our farm was now part of their life story in a way wholly unimagined by us, the guests of honor. We were so proud of our farm that day. That was a good day.

Another year, on the last day of September, our first farm dog, a Jack Russell terrier named Audrey, took her last, halting walk at dusk on the farm paths she knew so well (after surprising us with a full-throttled run at a flock of turkeys a few days before). Dog lovers will understand when we say there is one dog above all others in a life. Children always wanted to touch her wherever we went and they surrounded her at the farm table. They wanted to hold onto her and they didn't want to let her go. It was like that for us, too. Losing her brought us to our knees. Grief is always equal to its love. Her bright light flared out after sixteen and a half years, peacefully, naturally, after a slow decline, just as it should for any dog, or any loved one.

ABOVE Wedding tent. | Audrey.

OCTOBER

In which leaves change color . . .

Perhaps it is the sun's low angle or fall's early gold deepening to amber—there is a singular glow to this month above all others. The magic hour of twilight seems to last all day, and the farm is illuminated by its own brilliant light, at once captured and reflected. New patterns, contrasts, and relationships emerge as the leaves, grasses, and trees change shape and color. Pastel landscapes appear and disappear in the soft light. Grasses form into abstractions. Spectral mushrooms peek from under little tents of leaves and pine needles along the pathways, and milkweed seeds parachute past, dark seeds on glistening feathered wings. Pine cones become sculptured forms, and acorns and chestnuts, pocket talismans with blank faces and knobby caps. The netting posts are totem poles. And mugwort, that nasty invader, is transformed into a sea of lavender gilded by sudden splashes of light; it tosses and sweeps in undulating waves with every current of the air. It must be October.

Out in the field it seems as if a veil of filtered light is cast down gently each morning with sunrise and slowly lifted at the end of the day. We are moving toward the darkest day of the year, and in luminous October we have the last bright time of color before the somber months that circle ahead. There will be dramatic changes in the meadow and on the surrounding trees. What begins as a subtle shift in the blueberry rows will progress, slowly at first and then quicken. Over the next four weeks they will color together from deep green to a uniform burgundy. Green will linger longest on the lower branches as the wine color permeates and deepens. This coloration shift is due to the loss of chlorophyll, which slowly begins to disappear this month, first around the edges of the topmost leaves and then down the entire plant. As the green of chlorophyll fades, dappled yellow, orange, and red, the colors of other leaf constituents there all along, will become more visibly dominant. This kaleidoscope of color will ripple down through the rows and carry over into next month.

The warm waters of the nearby ocean may provide many mild and beautiful days in October, but the shortening days and cool night temperatures are sending signals to the plants. Soon, but not quite yet, they will slow down the growth of new shoots so they can "harden off" for the coming cold. The root system growth and the development of next spring's fruit buds will not slow down, however. Between the rows, the annual rye is an army of Lilliputian spears in peridot green protecting the blueberries from encroaching weeds. If necessary, we will irrigate this month if rainfall is spare. Weeding continues.

Fall grasses.
Milkweed seeds emerging. | Seed pod of the white campion wildflower.
RIGHT Turkey tracks on our dirt road.

The birds are quieter now. We notice this. We often come upon a dove or two walking quietly among the rows. There are many places for them to hide still under the canopy of the plants. As the weather chills, the hawk will hunt them and others more aggressively, but for now, on the warmest days, there is a peaceful coexistence. The raucous crows pass over on their daily patrols but seldom linger long, and occasionally a few swallows materialize out of thin air; the turkeys strut and toss fallen leaves about. The ever-present bluebird, shy and swift, is always focused on the hunt early in the morning and late in the day, hoping to bolster its slight reserves. As we go from barn to field and back again, there is an unmistakable scent that can only be called fall in the air—of pine, wood burning, and drying leaves. There is an invigorating chill. With dark, there is the full Hunter's Moon. When the moon is new or quartered, we can see that Orion, the Hunter, has already placed one foot on the sky's threshold. We can see the stars of his head, his arms, his tightly drawn bow reclaiming his place in the late evening skies.

October day.

Blueberry leaves take on russet tones. | Old barn doors.
BELOW Our collection of old fruit box labels.
OPPOSITE Brilliant fall light.

We have things to finish before our season is over—primarily mulching, cleaning equipment, and storing it away for the winter. Soon, we will be more often in the barn than in the fields. At one time, our barn, since moved, was situated on the road and housed our fruit and vegetable stand. Our greenhouses were used for growing seedlings. Now that we grow blueberries exclusively, both do well for storage and work areas, which any farm needs. Our larger equipment is here—the seeder, rototiller, sprayers, mower, and front-end loader. We store a tractor in each greenhouse along with two spreaders, the wooden trailer, and wagon. In the barn, various and sundry tools of the trade crowd the open space: an air pump, pruning equipment, shovels, rakes, pitchforks, mallets, a weed whacker. Spare irrigation lines gather dust in the rafters and boxes of spare nozzle heads sit on dusty worktables. Jars and cans are filled to the brim with hardware. There are rolls of netting, baskets of netting pins, spools of netting rope, boxes of pint containers, stacks of metal picking trays, barrels of replacement flags, and pails and buckets in every corner. A glass hurricane lantern is gauzy with cobwebs. Our sign lives here off-season, and above its resting place is a collection of vintage fruit box labels, tacked along the walls long ago. They glow in the dim light of the old barn amid the effluvia of a farmer's lifetime. And in the attic, a centuries-old steamer trunk from Galway is tucked under the eaves.

Ablaze with fall color.

As the monarch butterflies descend into Mexico at the end of their migration, they are thought to carry the souls of ancestors with them. So begins the Mexican celebration of Dîa de los Muertos, or Day of the Dead, on October 31. We will celebrate Halloween, or All Hallows Eve, the eve of the Christian All Saints Day. Samhain, celebrated on the same day, was, and is, the Celtic new year in pagan tradition. It marked the end of the summer's harvest with door-to-door trick-and-treating, as well as bonfires. The final cross-quarter day of the year, it lies between the fall equinox of September and the winter solstice of December. All seasonal rituals evolved to strengthen the bonds of relationship within a community and with the forces of the natural world, as life ascends in the spring, in birth, to descend in the fall, in death. During this liminal time of October, between seasons of bounty and hardship, light and dark, nothing is one thing or another. "Nothing gold can stay," Frost wrote in a poem of the same name, and no other month seems to express this polarity, of holding on and letting go, as strongly.

The grasshoppers do not leap and scatter when we pass now, and the chorus of crickets has fled. A lone chirp sounds here and there along the path like a clock running out or an old whistle left too long in the damp grass. Light emerges from the dark scrim of trees around the field; the setting sun baths the woods with a glorious backlit florescence. The change this month is a slow burning fire, soon down to its last blackened embers, holding the heat, flaring up and dying down. This is October.

Foxtail. | Milkweed seeds taking flight.

NOVEMBER

In which leaves fall . . .

The color of early November is well aged, a palette of deep, rich tones. What trees have lost their leaves only intensify the color that remains with their dark-fingered branches. Our blueberry plants will hold their leaves within their grasp until later in the month. In these thirty days they will undergo an enormous change, visually and botanically. We will see the darkest of reds and the brightest, sunny yellows, warm oranges, and lingering greens. When each leaf falls to the ground, it seems to do so with quiet surrender. There they will join the insulating leaf mulch we add this month and continue to provide nourishment throughout the winter as they decompose. But some enjoy a few weeks of freedom all their own. Leaves drift and spin from the branches of all the trees that surround the quiet rows. Some are as soothing to watch as snow falling. Others come down in a fury, pulled by a gust of wind. Once dried, they scurry merrily along the dirt road as if live things. Some whirl and spin in troupes of dervish dancers. We have listened to them fall at night, whispering past in the dark.

Speckled leaves.

We are finalizing our winter preparations, too, finishing up any chores left undone. Mulching is our highest priority. Every fall we keep our gate open to local landscapers who bring in the leaves and pine needles they've gathered. It's convenient for them, if they finish work late in the day when the dump is closed, to swing by our farm. A mountain of natural mulch rises up along the wood line in the far western corner of the field. We allow the leaves to sit for an entire year, forming moisture-retentive, nutrient-rich compost that is perfect mulch for our plants. We get to work up and down the thirty rows with our tractor and wagon while the mulch is still malleable and unfrozen. We usually have our first hard frost early this month. Once we are done, all of our machinery and tools will be thoroughly cleaned and oiled and stored away. Fortunately, Dave attends to much of the machinery maintenance and repair for us. Over the course of the year we sometimes lend equipment to other farmers, or borrow theirs, and we make sure everything is back where it belongs.

Switch grass with late fall colors.
OPPOSITE Row of blueberries with last leaves.

The USDA tells us, "A warm wet spring, favorable summer weather, and warm sunny fall days with cool nights should produce the most brilliant autumn colors." Things are rarely aligned so perfectly in nature. Leaf coloration at the farm depends on many factors. Weather conditions on Cape Cod do not generally conform to the rest of the state or its coastlines. Any one factor can change things from year to year. Every so often, all the leaves will be blown off early by a vagrant storm before most have had a chance to color, the plants stripped leafless almost overnight. One atypical and difficult year an Arctic air mass descended with snow and cold in November and lasted through late March—there was little in the way of color that season. By month's end, hurricane season is no longer a threat, but one year is now part of the farm's history: Hurricane Bob hit us unexpectedly in August 1991. Our nets were lost to the storm, blown into the surrounding woods, where we still find scraps of them clinging to tree limbs. The salt spray from the ocean storm surge coated the leaves, which dropped early that year.

Last month each bush blushed to a deep burgundy, from top to bottom, row upon row. Now all the bushes will take on their most vibrant color before they drop their leaves. The process of photosynthesis, active in the plants from May through early October, has been completely curtailed. When the plants were in their growth cycle, in early spring and fall, a small percentage of sugar was stored deep within their fibers for winter use. When weather changes signal the approach of winter, chlorophyll production, a vital component of photosynthesis, stopped and leaves lost their green color. Other pigments present in the leaves came to the fore, dappled at first and dominant later—the carotenoids, which are yellow, orange, and brown, and anthocyanin, the pigment that accounts for the berries' concentrated blue color and the canes' crimson tint in late fall. Now that life-sustaining fluids are no longer moving in and out of the leaves, the connection point between leaf and stem closes off and the leaves fall. Deciduous plants that drop their leaves have evolved this adaptation to conserve water in the plants and because leaves are simply too vulnerable in winter. As the blueberry plants lose their leaves, their fruit buds begin to harden off. Roots will continue to grow until the soil temperature goes below forty-six degrees.

In the meadow, wild grasses pale, the honey and amber of the early weeks will blanch before our eyes to a monochrome uniformity. A low mist often forms in the mornings as the temperature changes and the warmth rises into the cooling air. The last of the summer asters and goldenrod still beckon what few bees and butterflies linger. Bluestem glows in the waning light as wildflowers wither and brown and mushrooms darken in the woods. We hear the high-pitched whistle of the cardinal and the answering screech of the jays as they vie for wild berries. November has a feeling of inevitability, a momentum; a quiet, sometimes somber beauty that is unlike any other month. There are more gray days than sun-filled ones, and the sun sets earlier daily as we approach the winter solstice, shortened even more by the return to standard time.

The annual rye's now thickly matted corridors of green darken between the rows. The young canes of the blueberry plants turn deeper crimson, and the old canes' silver gray bark glows more luminously. The blueberries are poised to enter fully into their dormancy, or botanical hibernation, an incremental process that began last month. This restorative period marks the end of their months-long cycle of growth and the beginning of an inward turning that will last for at least three months.

The full moon this month is known as the Beaver Moon, no doubt for the industrious animal's purposeful winter preparations—for the way that he protects himself and his family from the threat of the winter months that lie ahead. Orion continues its ascent in the late evening sky. Though a beautiful constellation, there is an ominous quality to its appearance; the winter count once begun cannot be undone until he has had his full dominion over the night skies.

Rows of blueberries with winter mulch insulation.
BELOW Tractor with mulch spreader. | Mulching rows for winter.

It seems somehow fitting that the darkest hour the farm has ever faced should fall in this somber month. An act of wanton vandalism occurred one November day, undertaken with such reckless disregard for its consequences it is still troubling. A new neighbor, whose business backs on our farm, cut down nearly forty pine and oak trees of substantial size and age on a Sunday afternoon while we were out. We were not there to hear the whine and rasp of the chain saws and thud of falling trees. He and his crew clear-cut the trees down to stumps and uprooted all the undergrowth. He cut the logs into firewood, spread gravel and wood chips. His clear-cutting reduced the screening, provided by the trees, of his red metal industrial building. This exposure deprived us, and our patrons, of the pastoral natural setting for the farm that we had deliberately created and always enjoyed. The building was now an enormous eyesore visible from all the blueberry rows, our farm table, and most particularly from the dirt exit road.

The excuses he later offered, perhaps legitimate fears of vandalism and falling trees, held no corresponding concern for us—he never approached us with his concerns, nor did he ever offer any apologies. He acted, as many do on the Cape and elsewhere, to pursue his own interests without considering others, breaking

Mugwort in wildflower meadow.

the law by following the adage, "Better to ask for forgiveness than to ask for permission." What he did was wrong. It was also illegal. And it was tragic and heart-rending. Some of our evergreens and white and red oaks had stood there for one hundred years. Those that decay and fall are left in place to provide homes for wildlife and decomposing nutrients for the soil. His trespass and destruction on our farm was a stark reminder of the challenge we all face when others have opposing worldviews.

As of this writing the situation has not been completely resolved. We did act swiftly and reasonably to protect the farm and our rights; however, the situation was complicated by matters beyond our control and at times even beyond our comprehension. We had sold our eighteen-acre farm property to the town (using Land Bank funds) for its conservation protection ten years previous, retaining a life estate and exclusive use of the land as the prerequisite for the sale. The land's future use, upon our death, was to be as an open-space groundwater recharge area

The farm on a late fall day.
BELOW Bailey. | Trespass destruction.

In these final weeks of the year, all that is autumn fades and winter closes in over the farm. There will be no giving softness underfoot until the snows come. The dogs will stop their digging and we will don our warmest coats. The blueberry canes, shorn of leaves, will deepen into their richest shade of fiery red as the evergreens along the wood deepen into their darkest forest green. Soon enough the snow will fly. This month, dormancy will be fully entered. Days of bright light will be followed by sullen gray. What still felt soft and verdant in the landscape takes on sharper edges in these weeks as the days shorten.

In preparation for the snow and cold, the roots of the blueberry bushes are now bedded down beneath a thick, protective blanket of organic leaf mulch. Highbush blueberry plants do well up into the southern parts of Northern New England precisely because they can tolerate the climate here. In fact, it takes a certain ratio of cold, even frigid, days for them to fully enter into and then complete their dormancy. Once their chilling requirement, generally three to four months, has been achieved they will re-emerge, like all perennials, re-invigorated in the following spring. As the days approach their shortest length, the plants are now fully ready for the winter season. Carbohydrates have been stored in the roots to provide energy to resume growth in the spring. The water that remains deep within the plant cells has a high sugar concentration in order to lower the freezing point of water, acting as a natural antifreeze. The plants are now at rest in their full dormancy, using acclimating strategies developed over thousands of years. Where the older silvered canes have been thickened and weathered by time, the thin bark of the younger shoots is a vibrant, showy red; never is the contrast more noticeable than now.

The progression of dormancy is spread over several months, but once the blueberry plants have entered their fully dormant state, it becomes evident to the eye. It is not simply that all the leaves have been lost, but that the branches have turned a deep shade of red. It seems remarkable that weather can be as variable as it often is from September through December, and yet their transformation occurs on schedule every year, and in every season. The blueberry plants held to their steady progress through the stages of dormancy throughout the fall. They read the signs of seasonal change along the fibrous bark of their skin; they read the lack of warmth, though not yet frigidly cold; the hardening of the ground, though not yet completely frozen; the shortening hours of daylight. The blueberry bushes, not unlike the other flora and fauna in the field, prepare, and they are not caught unaware. When the berries ceased production in early September, and new growth ceased in October, the process of withdrawal was still an undercurrent; it is fully realized now. Temperate weather will not effect any change in their internal clocks.

OPPOSITE Netting posts stand sentry.

If we have not gotten to it already, we finalize our own preparations. We always take the time, tedious as it may be, to ensure that all the irrigation lines are blown free of debris and standing water. This can be a chore—one we put off, but it is an essential part of winterizing the field. By this time the tractors and other farm equipment will have been put up for the winter after being thoroughly checked over. At the close of every year, we tally our yield from the summer into log books we have kept for decades now. We do keep a daily tally during the summer, but this is when we get a chance to compare the season to others. We have always tried to keep to a middle course between expense and profit, as any business does, within a variable margin of occasional highs and lows. Most years the log books show a wholesome balance between expenditure and yield, though rarely profit, and it is a steady enough course to keep us happily in business.

The winter moth returns. Having made its first appearance at the end of November, this month they flit freely about among the trees in large numbers. The moth has emerged out of its pupal stage under the soil, where it lay dormant since late spring. The moth will mate until the temperature is consistently below freezing, usually in January. Only the males fly; females await their discovery at the base of plants. These are nocturnal insects, and there are few predators at this time of year, especially at night. The males are attracted to light, and swarms can be seen fluttering around lampposts and in car headlights. It is a large yellow moth, soft to the touch, with a powdery surface texture. It may be an attractive and novel sight to see in winter, but if the population flourishes now it will do its damage come spring. The winter moth is very destructive to certain trees and shrubs, including blueberries, when its eggs, laid now, hatch out.

It is the time of the Full Cold Moon and Orion's grand procession across the velvety midnight blue sky. The earliest sunset has already occurred around the 6th of December, as there is a disparity between our timekeeping and the sun's, but these are still the darkest weeks of the year. The winter solstice on December 21, marks the shortest day, with only nine hours and five minutes of daylight compared with summer's high of fifteen. The shortest day means the longest night, and at mid-heaven at nine p.m. in mid-December the hourglass figure of our winter constellation Orion has reached the apex of his ascent. He is once more, for a brief time, the central radiant figure in the revolving nightscape, his hourglass full and still. Is it any wonder that light is at the center of all mid-winter's traditions in these silent nights?

All of the remaining leaves will have fallen to the earth, having lost by several months, irretrievably, the sustaining hours of sunlight and nourishing sap that fostered their growth. Each blueberry bush has drawn its resources deep within. The full circle of the year is complete. In these short days, the landscape seems to have adorned itself in yuletide spirit—the blueberries with their red canes flanked by the forest-green spires of the pitch pine, our few holly trees adorned with cardinal-red berries, striped red-berried wintergreen underfoot, and blue-

Crimson canes. | Winter wildflowers in golden light.

Snow moving in.

green juniper berries clustered into little beaded swags. The rush of the winter wind in the tall pines and bare oaks encircling the field can be mistaken now for waves rushing headlong against the shore.

Few sights are as stunning as bare black branches against the firelight glow of the setting sun. In that twilight hour, the delicate-fingered tree limbs hold the twinkling light of emerging stars as if they were precious jewels, small diamonds. These many trees of life, where would we be without them? They take our breath away as they give us breath. These woods, meadow, and fields have fed us, body and soul, all year. And as the year closes, the seeds have already been sown for the next. So the years pass. We mark and name their passing.

A few words of reflection on the moon before we close. It has been our constant companion through the year and through these pages. We have marked its fullness with names. But it is not only its fullness that we note. As it moves through each of its phases, from new to crescent to full and back, waxing and waning, we are reminded of something closer to Earth. In its constancy and change we see something of our farm and of ourselves. We are all held fast by the sun's gravity, as we hold the moon, through the months and the years. Its rather wise and benevolent face is always turned toward us. Indeed, it may be part of us torn away and held close. Its force moves the seas in their daily tides, and it moves the sea of our bodies and the water hidden in the earth. The moon also stabilizes our

The circle of the year closes as it began.

climate. There would be no higher life forms on Earth without the moon. It is a mere 238,000 miles away, whereas the nearest star, Alpha Centauri, is 4.4 light years away (about 5.88 trillion miles). The moon is no romantic after-thought, it is as essential as the sun. It reminds us that what was full must empty and what was empty must fill. That is the cycle of the year and of our lives. And if we look to it to enlighten us, Zen master Dogen's moon in a dewdrop finds its equivalence in Blake's infinity in a grain of sand. Nature is our greatest teacher, inviting us to see with new eyes, showing us certainties and eternities beyond the constraints of our limited views, more and more deeply so with each passing year.

Blue moons. They say that a blue moon, which occurs every 2.7 years (a second full moon in a month) brings good fortune, as does any rare occurrence, such as a heron or a four-leaf clover. It could just be that the lunar cycle and our calendar are not in perfect accord. Beliefs are a matter of personal choice. We choose to think of them as fortuitous. We have calculated that there have been more than twelve blue moons on this farm over the forty years we have been here—a full cycle of blue moons. In fact, we met on one of them.

At the close of every year, we are grateful for the commitment this land has called from us to protect and preserve its integrity. It has given us our lives as we have given our lives to it. And we consider the blueberries—they are the fruit of all our labor. They give themselves to us so freely, sustaining us and so many others. These precious, perfect, blue spheres with their five-pointed stars are a little bit of heaven on Earth.

It is time for their winter rest and ours. As the first snow falls, we are safe and warm within, the hearth fire lit, the candles aglow, the dogs at our side. We reach for those books we love now, the ones that are time-tested and true, that speak to our hearts, and read this, as we do every year:

"I said some words to the close and holy darkness, and then I slept" (Dylan Thomas, *A Child's Christmas in Wales*).

PART TWO | The Histories

Elizabeth White and Frederick Coville inspecting hybrid blueberry bush, c. 1920.
Courtesy of the Whitesbog Preservation Trust.

A BRIEF HISTORY OF BLUEBERRY CULTIVATION

The Botanist and the Businesswoman

Frederick Vernon Coville, an American botanist, is credited with establishing blueberry cultivation, along with Elizabeth Coleman White, a businesswoman whose family owned an expansive cranberry bog. Coville was born in 1867, in Preston, New York, and went on to become the chief botanist of the United States Department of Agriculture (USDA). He was the first director of the United States National Arboretum. A graduate of Cornell University, he taught botany before joining the USDA, where he would remain until his death in 1937, publishing approximately 170 papers and books and participating in many botanical research expeditions. He began his research into blueberries in 1910, discovering the significance of soil acidity to the plant's success. He also conducted experiments on temperature, pollination, and plant selection from the wild on his home farm in New Hampshire. He collected his findings in monographs for the USDA—"Experiments in Blueberry Culture" in 1910, followed by "Directions for Blueberry Culture" in 1921. He wrote that blueberry cultivation could have a "profound significance" on agriculture and the general welfare of individuals in that the crop grows best in "worthless" soil.

Coville was the first to successfully cross wild lowbush and highbush blueberries. His Bluecrop, Blueray, and Earli-Blue varieties are still popular today. For this work he received the George Roberts White Medal of Honor from the Massachusetts Horticultural Society.

Elizabeth Coleman White was born in 1871. She was ten when her family settled in New Lisbon, New Jersey, purchasing a large estate with a cranberry bog, which they named Whitesbog. White would live there her entire life. She began to work at her family's cranberry bogs when she was twenty-two, and was an adept business partner for her father.

OPPOSITE BELOW US Department of Agriculture Pomological Watercolor Collection. Rare and Special Collections, National Agricultural Library, Beltsville, MD 20705.

Upon reading Coville's 1910 publication, White, with her father's support, wrote to the botanist and invited him to Whitesbog. She suggested that it might be an ideal place to continue his research. Given the acidic nature of pine barrens soil, the abundant presence of wild varieties of blueberries, and White's enthusiasm, he agreed. Their partnership lasted five years but its effects have been long-lasting. Both in the prime of their early forties, they made ideal partners—she a dedicated agriculturalist with strong business acumen, and he with a dedication to botanical science. White understood that the locals knew where to find the best wild blueberries and appealed to them for help. They identified the choice bushes for propagation, were paid for their efforts, and had bushes named after them. Thousands of cuttings later, in 1916, White and Coville produced their first commercial crop. White designed containers and wrappings that made the product attractive and ready for market, and they began shipping them nationwide. Blueberry cultivation and a new agricultural industry were born, not only nationally but globally; one hundred years later, blueberries are now cultivated all over the world.

White was one of the founders of the Blueberry Growers Cooperative Association (later known as Tru-Blu) and was the first female member and president of the American Cranberry Growers Association. She received a citation from the New Jersey Department of Agriculture and numerous awards from horticultural societies across the country. White was disappointed when her father willed the business to her brother-in-law, wanting a man at the helm of the business. Family members did not support her interest in diversifying the company, so she built her own home on the property at age fifty-two. White created an extensive garden of native plants at her home and founded a nursery business independent from the family, growing holly trees. In her later years she was active in increasing awareness of the welfare of migrant workers. She spoke widely, and was fondly known as the "Blueberry Queen." She died in 1954. Whitesbog is still active today.

Note: The illustration on the title page of this section was done by artist James Marion Shull (1872–1948). It is part of the Rare and Special Collections of the National Agricultural Library in Beltsville, Maryland. The artist "collected in the wild" various blueberry specimens at Whitesbog from which he rendered his exquisite watercolors in 1940.

A BRIEF HISTORY OF
Farming on Cape Cod

From the Brewster Historical Society Archives. Dennis farmers with hay wagon, c. 1900.
Photograph by Carol A. Dugan.

When winters here are their coldest, the seas themselves freeze and great ice sculptures are heaved onto the beach. As far as the eye can see, it is all gray ice. When the last ice age held the continent in its grip 25,000 years ago, there was no land here at all, only ice. We can almost imagine this on a cold, wintry day; it makes us shudder to see no sign of life anywhere. Cape Cod was formed at the retreat of the last glacier, the Laurentide, which marked the end of the last ice age.

By 13,000 years ago the glacier had retreated elsewhere, taking with it the Bering Land Bridge from Asia to North America. The sand, gravel, and rocks left in the retreating glacier's wake, acted upon by the tides and rising sea levels, built this peninsula as an outwash plain. As the glacial ice melted, fresh water was released and plant life began to take hold in the shallow, sandy soil, strewn with rocks. The archaeological record shows that pollen from established trees was present 12,000 years ago. Grass and lichen had most likely taken hold long before. Nomadic native tribes began to frequent the area and eventually establish themselves, calling the place "This Narrow Land." Cape Cod covers 339 square miles, extends sixty-five miles into the Atlantic, and averages five miles across, from Cape Cod Bay to Nantucket Sound. Climatologists predict that as climate change takes hold in the centuries to come, and sea levels rise, this fragile coastline may disappear beneath the waves once more.

The Paleoindians who roamed in search of game, fish, and wild edible plants were later known as the Nauset and Wampanoag tribes, part of the extended Algonquin Nation. They moved and camped with the seasons, building portable structures known as long houses for extended stays. Corn, beans, and squash were their first crops. These vegetables are known as the three sisters. Research undertaken by botanists, geneticists, and archaeologists has confirmed their origination in Mesoamerica. Corn, or maize, was domesticated from the wild rye grass, teosinte, 9,000 years ago in a process of cultivation that may have taken the early farmers a millennium to perfect. These first ingenious botanists and farmers shared the secrets of their hard-won cultivation practice and it spread across the continent. It was not until 1,000 years ago that any archaeological evidence points to settlement and its attendant agriculture on Cape Cod, though sea and sand have buried much over the millennium.

Native Americans harvested wild blueberries, "star berries," along with cranberries and Concord grapes. These are America's three native fruits. They also foraged for acorns, mushrooms, tubers, and various edible plants, some with medicinal value, such as dandelion, mugwort, and willow. Pemmican, a dried fruit and meat jerky mixed with stone-ground corn, extended the harvest season. These early hunter-gatherers and fledgling agriculturalists created weirs, a series of stone and wood dams, to trap fish, which they caught with spears. Especially plentiful were alewives, river herring, that came in from the sea every spring to spawn in the inland ponds. Herring was also used as a fertilizer for their corn. They cut trees to create farmable land. After five years of use the area was burned and the soil left fallow to recover. Corn was a staple crop that provided an abundant harvest, as it could be dried and stored underground in mounds over the winter. Over time, corn, beans, and squash harvests were supplemented with pumpkins, tobacco, gourds, and eventually melons and artichokes.

The first European colonists arrived on the Cape aboard the Mayflower in November 1620, landing on the tip of the peninsula, Provincetown today. They were a separatist group of Puritans from England, known as Pilgrims. Mayflower voyager William Bradford, future governor of Plymouth County, wrote that the land was "all wooded with oaks, pines, sassafras, juniper, birch, holly, some ash, (and) walnut" virtually to the sea. The beauty of the sheltered harbor, the sandy beach and the deep woods, its profound quiet and dark nights, must have been powerful. Enlivening perhaps after being so long at sea, but no doubt a frightening aspect on a wintry November day. Bradford lost his wife, who fell overboard and drowned in the harbor before they set sail to scout the area, eventually settling in Plymouth.

The colonists survived the winter with the help of the local Indians and farmed communally in their first spring and summer. Within a short period of time each resident was granted an acre of his own land to till. Planting was hard work, done by hand with crude implements. Farm help grew as families expanded. Quite unexpectedly, the new settlers were the beneficiaries of a cache of open fertile land. In the period from 1616 to 1619, seventy-five percent of the Native American population had perished, decimated by European diseases they had no immunity to. The settlers inherited the rich farming land they left behind. After a decade in the new world, small family farms had been established with more substantial acreage and common pasture land for the livestock that was beginning to arrive from England. There was enough surplus to trade with natives and to begin to work off the debt they incurred for their passage. The settlers' villages were no more than clusters of family homesteads, modeled on the English country style, with a small home and kitchen garden. They were herbalists by necessity. A barn, coop, and fruit orchard extended beyond the yard, with common pasture land extending beyond the private boundary lines of each family's allotment. The fair use of pastures, as well as shared woods for tree cutting, were overseen by local township committees. The staple crops were still Indian corn, squash, and beans, to which were added rye, wheat, cucumbers, radishes, and cabbage, and eventually, potatoes and onions. The orchards they planted soon supplied them with apples, pears, and cherries. Hunting, fishing, and seasonal gathering of nuts and berries, augmented by seasonal migrations of fowl, geese, ducks, and river herring, provided stable sources of sustenance.

Harvesting on the farm was small-scale at first, as tools were rudimentary—hoe, scythe, sickle, and axe. With the increase in livestock and animal husbandry, change accelerated. Horses and oxen on the farms meant the ability to clear, till, and harvest with more sophisticated ploughs. Blacksmiths and craftsmen began to settle in villages. Harvests increased enough for farmers to sell their surplus goods, some at harbor markets.

After fifty years of fairly peaceful coexistence, things changed with the Native community. Wampanoag chief Massasoit may have befriended and assisted the first arrivals, but once he died his son, Metacomet (King Philipp), turned against the ever-multiplying settlers for a variety of reasons. King Philipp's War took place between 1675 and 1678, in the Bay Colony. Lives were lost on both sides, but the outcome was always in favor of the colonists; Natives numbered in the low hundreds, the settlers over 30,000.

At the turn of the seventeenth century, the first generation of native-born sons and daughters of the first settlers were in a better position in the New World than their forebears had been. As the first generation passed and families and villages grew, governance of common lands became unmanageable and individual ownership took its place. This was a significant but necessary change for the early settlers, but a dramatic shift in world view from the land's original Native Americans. Agriculture and animal husbandry grew exponentially as more and more settlers arrived, many experienced farmers. More homes were built and tracts of land were acquired, divided by stone walls, fences, and gates. Increased demands were placed on natural resources to support the growing communities.

Enormous changes in the ecology of the Cape took place in the eighteenth century. It may not have supported the expansive agrarian lifestyle of interior Massachusetts, but the land was arable enough for certain crops. Abundant springs and numerous freshwater ponds, approximately 365 kettle holes left by glacial action, provided plenty of water. The inter-tidal zones of extensive marshland, which changed the tides twice a day, made for nutrient-rich salt hay for livestock. Surrounded on three sides by the ocean, marine life was a mainstay. Each farm owned at least one pig, a cow, a few sheep, a goat, and some chickens. Apiaries were added to the farmstead as bees were imported from England to increase pollination and provide honey. No doubt cats and hardworking dogs, as well, joined the farm family. One long, winding dirt road, well-traveled, with stage coach stops that evolved into taverns, ran from the head of the Cape to its tip, Plymouth County to Barnstable County, and most villages were situated off this Old King's Highway, now known as Route 6A.

By the early 1700s, grist mills had been built on local streams, the coursing water providing power to turn the enormous wooden wheels used to grind corn, thereby reducing farm labor. Windmills were erected in open fields. The increasing abundance of corn, and now cornmeal, led to increased trade and strengthened the local economy as goods, as well as people and letters, were transported to Boston Harbor by small sailboats, called packets, that put out from Cape harbors. An economy beyond the family farm, based on agriculture and fishing, was quickly evolving. By the mid-eighteenth century, farmers turned an eye toward exportation of more than corn, to salt hay, horses, cattle, and sheep. Fishermen added whales to their catch, creating a new, thriving industry. More wood was needed for building new ships for fishing, whaling, and transport, and these new ships would need provisioning. Caught up as the settlers were in establishing themselves in this new world, there was little time or interest in taking measure of the land's sustainability for all their enterprises into the future.

Young men did leave the farm to join militia during the Revolutionary War, from 1775 to 1783. Of those who survived, some would never return, seeing better opportunities in trade, law, politics, or higher education. Imported items were scarce during the war years, and an enterprising farmer, John Sears, of Dennis, began to experiment with evaporating salt out of seawater. At first he used wood fires, but then turned to solar drying. His method was patented in 1799, and for nearly one hundred years, the salt works of Cape Cod provided salt for homeowners and fishing fleets, preserving catch for maritime trade and preserving provisions for sailors months at sea. By the 1830s there were 442 salt work operations powered by small windmills throughout the Cape.

With the Revolution behind them and industry increasing, agricultural clubs and societies began to form in the agricultural communities. Farmers and gardeners had long kept journals or daybooks to guide them from year to year. Societies shared anecdotes and tips. In 1792, this practice was popularized and extended by the publication of the *Old Farmer's Almanac* out of New Hampshire, the oldest continually published periodical in America. The popular handbook was replete with astronomical information, uncannily accurate weather predictions, planting charts, and tips, games, recipes, and folklore. It even came complete with a hole to hang from a nail in the barn or farm kitchen, ever at the ready.

The landscape of Cape Cod had undergone a dramatic change in 200 years, as had its sheltered harbors, now bustling with cargo ships, fishing vessels, packets, and schooners. In 1815, the first schooner was built on Cape Cod at the Shiverick Shipyard in Dennis. No longer did sea captains have to put out to sea from Boston. These long-distance voyagers brought great material wealth back to the Cape with them, building the first mansions along the main road. Their homes, though situated on extensive farmlands, were clearly the homes of prosperous merchants, not farmers. Families and communities grew at the expense of a diminished pastoral landscape. Family farms grew more and more surplus crop for export. And cultivation of wild cranberries had just begun in 1816 in Dennis. Captain Henry Hall, a Revolutionary War veteran, noticed that his wild cranberries grew better after a windstorm had blown sand over them. He copied the technique, transplanting wild plants into his bog and covering them with sand. They did grow better, and a commercial cultivated cranberry industry soon flourished across Cape Cod and continues to this day.

As the nineteenth century advanced, farmers found new markets for their excess goods by the bustling harbors. Wheat, onions, turnips, cranberries, fruits, cereals, vegetables, fish, and oysters were all being purchased and shipped off Cape on packets. The Barnstable County Agricultural Society held its first agricultural fair in October 1844, at the County Court House in Barnstable Village, with exhibits, horse shows, livestock exhibits, fireworks, and food. A fraternal organization known as the Grange came into being soon after, in 1867. Communities opened their own chapters to support their farm families and communities. The Grange was the first agricultural advocacy group in the country, and through its efforts free mail delivery was instituted, from which everyone, not only rural communities, benefited.

The United States Department of Agriculture (USDA) had come into existence by this time, created in 1862, by President Abraham Lincoln. He called it "The People's Department," as fifty percent of Americans lived on farms. Born and raised on one himself, he understood their particular hardships with the keenest empathy. Farmers needed the recognition and support of their government.

American poets gave voice to the land through the works of Walt Whitman, 1819–1892; Emily Dickinson, 1830–1886; and later, Robert Frost, 1874–1963.

There were consequences to the region's expansive growth, and the early residents of Cape Cod would look fondly back at what was called "The Golden Age" from 1790 to 1850. By 1849, when Thoreau made the first of four trips across the Cape, he called it a "vast and desolate place." Notable to him was the absence of trees and a landscape that seemed more wind-scoured and arid than fertile. Indeed, as early as the mid-nineteenth century, the larger Cape-made vessels were being towed to Boston for rigging, as the Cape woods could no longer supply the Dennis shipyards. Trees had been felled to clear land for farming, for wood to build homes and barns, to make furniture and provide heat, for building ships, and for export.

After one hundred years, the supply no longer seemed endless. Deforestation and its ensuing erosion began to be problematic at the onset of the nineteenth century, and by mid-century, hard times were prevalent for Cape farmers. Not even the whaling industry could support itself. The population in 1800, had been 19,000; in 1850, it was 34,000. The railroad, which had reached Sandwich in 1848, laid tracks all the way to Provincetown by 1873, bringing a rise in tourism. Farmers adapted their market once again. New hotels needed to be supplied with food. Seeing the importance of scenic vistas, towns would appoint tree wardens, and trees would be planted along every roadside.

At a national level, the American Farm Bureau Federation, known simply as the Farm Bureau, a non-profit organization, was established in 1911, "to enhance and strengthen the lives of rural Americans and to build strong, prosperous agricultural communities." The Cape and Islands Farm Bureau would soon follow. In 1914, the USDA created the Cooperative Extension Service. Each county of each state had its own office and representative. The Extension was modeled after agricultural clubs and societies to provide educational, research, and general support on issues affecting rural and agricultural communities. The USDA partnered with land grant universities; it was a federal effort but managed through an educational body. On the Cape, that educational body is the University of Massachusetts, a land grant college since its 1863 inception as the Massachusetts Agricultural College. The Woods Hole Oceanographic Institute offers marine-related support through this partnership. The local extension office provides farmers with a dedicated entomologist, horticultural expertise, Buy Fresh Buy Local public relations efforts, soil and water testing, educational talks, and programs such as master gardener certification, among other things. The USDA's Natural Resources Conservation Service, with local offices nationwide, provides technical assistance and grants to local farmers, cranberry growers, aqua-culturists, and wetlands restoration efforts.

In 1888, the Cranberry Growers' Association was founded, and within a few short years, 3,500 acres of cultivated cranberry bogs could be counted Cape-wide. In 1892, the Massachusetts Dairy Bureau was established. Four years later the

Massachusetts Audubon Society followed, an organization that would have a lasting effect on local conservation.

During WWI, in 1916, the Cape & Islands Farm Bureau was established to support and oversee the changing face of local farming. In the early part of the twentieth century, Cape farmers were focused on crops that flourished best in the often challenging soil. In its natural state, most Cape soil is sandy, acidic, and lacking in fertility, but some areas have prime farmland. The local crops that performed consistently well included turnips, beets, cranberries, onions, potatoes, asparagus, apples, strawberries, and blueberries, which would be cultivated after 1920. That same year, Falmouth was one of the largest strawberry producers in the US.

Dramatic changes were ahead for Cape farmers. In 1921, the Cape Cod Chamber of Commerce was created. The railroad was now carrying passengers all the way to the tip of the Cape, especially during the summer months. Demands on land use for cottages, hotels, and tourist-related industries started to compete with the open space needed for farming. Land was sold, and a smaller-scale agricultural industry that catered to the burgeoning tourist trade evolved. In 1920, 47,000 acres were dedicated to farming. This amounted to eighteen percent of the land. Most farms were ten to twenty acres, and the most successful had roadside stands. These were part-time farmers, as most needed other jobs to survive. The Barnstable Municipal Airport opened in 1928. Fields, woods, and dirt trails were paved into roads, parking lots, and runways. By 1940, only fourteen percent of land was being farmed, and by 1950, only eight percent. As tractors steadily replaced horse and mule labor on farms after 1946, productivity increased exponentially. Although large-scale industrial farms benefited most from this technological revolution, Cape farmers enjoyed the ease of tractors as well. Horse power of a non-mechanical nature became a thing of the past.

After WWII, the 1950s were a time of optimism and prosperity for the middle class, who now took family vacations in cars. Trains discontinued their service in 1959, after the Mid Cape Highway was opened, ushering in even more local development. In 1950, tourism accounted for seventy million dollars in the local economy, whereas fishing, agriculture, and local manufacturing concerns provided only eight million. Cranberry bogs were filled in and land was sold off. Farmland and developable land began to compete with each other, and some farmers who could not sustain themselves with a family farm and part-time work subdivided and sold their farms. Only seven percent of the land on Cape Cod was developed in 1951, but by 1997, that figure had jumped to thirty-eight percent.

Protecting the environment became a driving force after the Environmental Protection Agency was established in 1970. One local impact was wetlands regulation. Tidal estuaries—which had been blocked by dikes to create new pasture land, keep the mosquito population under control and provide more acreage for development—came under state and federal protection. The Association for the Preservation of Cape Cod was created in 1968, and in 2010, the Compact of Cape

Cod Conservation Trust was formed to preserve open space. It would take until 2007, for the remnants of the Wampanoag Nation on the Cape Cod mainland to be granted federal recognition as a tribe with a small tract of land in Mashpee. Most of the Cape's rich woodlands are gone. Re-forestation consisted mainly of scrub pine and oak in fallow fields, and small wood lots rather than the diversity of old tree species, though some maple, tupelo, birch, and beech are still present. Vistas today are mostly of marshlands. Most beaches, outside Cape Cod National Seashore's 43,607 acres of beach, dune, and forest, are lined with homes, as are lakes, ponds, and kettle holes. There are few fields, meadows, or farmlands in easy view.

According to the Compact of Cape Cod Land Trust, as of 2016, forty percent of Cape Cod is developed, and forty-four percent is under conservation protection, including the National Seashore fortuitously protected by President Kennedy in 1961. This means that only sixteen percent is still available land, for either development or conservation. According to the US Census Bureau, as of 2000, the Cape's population was 222,000, with an additional 300,000 tourists during the summer months. Just forty years previous the population had been 70,000. In 1930, the population density per square mile was eighty-two people; by 2000, it was 562. According to the USDA-NASS (National Agricultural Statistics Service), in 1925, there were nearly 9,000 acres dedicated to agriculture; in 2007, only 2,000.

There are approximately 476 farms on Cape Cod today, making up 4,269 acres of active farmland. The average farm is small but intensely cultivated. There isn't much of anything they do not grow. It is worth repeating that only 150 years ago fifty percent of Americans lived on farms, and today only two percent do. But today people are increasingly concerned with the quality of the food they eat and small farms are growing in popularity. The most profitable local crop is cranberries.

The USDA census definition of a farm is any place from which $1,000 or more of agricultural products were produced and sold during a census year. According to the 2012 USDA statistics, there are 7,755 farms in the state of Massachusetts, covering 523,000 acres. This is the sixth highest number in the US; eighty percent of these farms are family owned; ninety-five percent are small farms; the average farmer is fifty-eight years old. Female farmers represent only thirty-two percent, but this is an increase from previous census years. The farm is the residence for seventy-five percent of farmers, but the farm does not provide enough income to support the farmer without additional off-farm work.

Those 7,755 farms in Massachusetts, on 523,000 acres, produce $492 million of revenue. The state ranks third in the nation for direct market sales per farm and sixth for Community Supported Agriculture (CSA), which has increased by a dramatic ninety-five percent since 2007. Agri-tourism sales increased over the same period by a staggering 127 percent. There are 409 blueberry farms in Massachusetts, covering 773 acres. Maine has the highest number of wild blueberry farms, at 510 farms, but has only 290 cultivated blueberry farms. The top five blueberry-producing states in the US are, in order: Washington, Georgia, Michigan, Oregon, and New Jersey.

Locally, farmer's markets are on the increase, along with agricultural initiatives such as Farm to School, Buy Fresh Buy Local, town-operated community gardens, school gardens, CSAs, farm stands, aquaculture leases, fishermen alliances, agri-tourism ventures, and bee-keepers associations. Cape farmers continue to benefit from many supportive agencies, such as the USDA-NRCS, the Southeastern Massachusetts Agricultural Partnership (SEAMASS), the Massachusetts Farm Bureau Federation, the Cape & Islands Farm Bureau, the Barnstable County Cooperative Extension Service, and the Cape Cod Conservation District. The county fair tradition continues in various towns each fall, and every spring for the past seven years the Cape and Islands Farm Bureau and a local conservation trust host a Meet Your Farmer Day. Continued support for our small local farms must continue. There is little land left. Conservationists can often oppose the use of agriculture on land they would rather see as open space, and compromises will be necessary. Towns will need to enforce protection of agricultural by-laws against the complaints of landowners who do not want farms in residential neighborhoods.

A BRIEF HISTORY OF
Hokum Rock Farm

Hokum Rock Farm c. 1988.
Photos by Stephen Spear and Ed Harper.

In 1944, when he was on leave from the service, my dad bought eleven acres of land with a small cottage in East Dennis, Massachusetts. When he got out of the service, he attended college on the GI bill like so many others. He returned to the Cape and worked as a science teacher during the school year and a chef during the summer. He had no farming background, having been raised in a factory town in Connecticut. He married my mother in 1956. He was actually my stepfather, raising me from the age of five. We lived in the cottage for a year and then moved off Cape in search of a better teaching opportunity. We spent time in the cottage off and on during the summers that followed and lived mostly in Connecticut.

I lived in the cottage while working summer jobs during my college years. After I graduated in 1972, with a degree in English, I was back on the Cape working as a carpenter's helper. I was trying to make a decision about my future and whether to go back to school for a graduate degree. My parents were visiting, and Dad and I were planting a garden as we had done in previous summers. By this time, he had acquired a university position teaching botany. He brought with him that summer a flat of about fifty tomato plants that he'd used in one of his classes.

We only needed a few of them for our family garden. Like a lot of young people do, I was complaining about having to work for someone else and said that I'd like to work for myself. He pointed at the rest of the plants and said, "Why not plant them all and sell tomatoes by the side of the road?" His suggestion was made half in jest; however, I took the idea seriously and planted all the plants, having no idea what would happen.

That summer I made a grand total of $27 selling tomatoes. Good thing I kept my day job! I learned something important, though. I loved being outside and working with plants and wanted to keep doing it somehow. By fall, my parents and I began to develop a plan to start a farm. For me, it would be a way to earn a living; for my parents, it would be a retirement activity. As a botanist, my dad loved plants, and my mom looked forward to the retail side of the operation, as she enjoyed working with the public, having done so in a number of different jobs. So a family enterprise was born, much to our surprise.

At the end of that summer, Dad noticed an article mentioning a man named J. Herbert Alexander. He owned a nursery about fifty miles away in Middleboro, Massachusetts. He was a well-respected breeder of blueberries and lilacs. He was also a distant cousin of my dad's, who had not seen him since childhood. When we went to visit him, we found him on his hands and knees weeding his blueberry plants. He was more than eighty years old. We spent the next two hours helping him weed while he told us how to start a blueberry farm. Among other things, we found out that it was a long-term, expensive investment. If all goes well, it takes about six to eight years after planting to begin to make a profit. We bought forty plants from him that day. It was a small start, but a beginning nonetheless.

A field of trellis tomatoes.

Vegetable plant seedlings in the greenhouse.
BELOW Chet and Selma planting seedlings.

Dad and I realized that we would have to grow other crops that matured faster in order to raise enough money to plant an entire field of blueberries. At the time, we only had a quarter-acre of grassland that could be quickly and easily planted. The rest of our land was wooded. I spent that winter and the next clearing the land. It was time-consuming and expensive. In the meantime, we planted that quarter-acre with vegetables, selling our crop off tables by the side of the road. After two years I had managed to clear an acre of land, and we built a small barn to use as our farm stand. We also purchased fruits and vegetables that we did not grow in order to have a wider variety of produce to sell. That year, I went back to college at the University of Massachusetts to get a second bachelor's degree, this time in plant and soil science. Two years later, when the next acre was cleared, we planted strawberries, which we sold on a "pick your own" basis.

Our business grew rapidly. We ended up with five acres under cultivation—strawberries and ten different vegetables. We hired ten people in the summer to help with planting, harvesting, and selling at the farm stand. Our experience was not unlike other new farmers or anyone else starting a business. A lot of money was coming in, and almost all of it was going out for labor costs, equipment purchases, and other expenses. Some winters I took other jobs to make ends meet. It was hard to put much money aside to develop the blueberry field. Finally, in the winter of 1981–82, I began to clear another field to plant blueberries. In July of 1982, my dad died suddenly of a heart attack. It was difficult, but my mom and I got through that season with the help of friends and great employees.

Pickers in the big vegetable field.

HOLLAND TRANSPLANTER
HOLLAND TRANSPLANTER

OPPOSITE Planting strawberries. | ABOVE The barn produce stand.
Strawberry harvest. | Various harvested vegetables on the truck going to the stand.

In 1983, I was offered a winter job working on a special project with the local Natural Resources Conservation Service (NRCS) office (then called the Soil Conservation Service). This agency, part of the USDA, provides technical advice to farmers and other large landowners to help them conserve their land's natural resources, which benefits everyone. I really liked the people in the agency and believed in their mission of having "a productive nation in harmony with a quality environment." I worked for them again the following winter, and when they offered me a part-time, year-round job I accepted it. The schedule was flexible enough that I could still farm, but not in the way that I had been. I would not be able to stay in the vegetable and strawberry business, develop the blueberry field, and also work for NRCS. There would not have been the time to do both properly. This opportunity prompted me to make an important choice. I decided to get out of the vegetable and strawberry business. The simplicity of growing just one crop made it practical to work off the farm. I sold off part of the farm and used the money to pay off my loans and establish my blueberry field. In 1986, with the help of an employee, I finished planting three acres of the crop that my dad and I most wanted to grow.

For the next three years, I followed standard practice of pruning off the fruit buds so no berries would form. This allowed the plant to put its full energy into growing strong roots and branches. In 1989, I invited the public in to pick the small, first-year crop. My mom died later that year after a long illness but was happy to see that first harvest. The public picking continued over the next few years, but I was not pleased with how the plants were growing. None of the diseases or insects described in professional blueberry growing guides seemed to be the problem. In October 1991, a clue appeared. The wood chip mulch that I used around the plants had always had various mushrooms growing on it. But this time I saw mushrooms growing from the ground right in the middle of the blueberry plant. Just to be on the safe side, I took some of the mushrooms to the local Cooperative Extension Office plant pathologist for identification. He keyed them out, then sent the information to a forest disease specialist at the US Forest Service, who confirmed that it was, in fact, a fungus called *Armillaria mellia*. Most common in the forest industry, it is often called oak root disease, but that is misleading because the fungus attacks a wide variety of woody trees and shrubs all over the world. In addition to blueberries, it infects other small fruits, like raspberries and grapes, most tree fruits, many ornamentals, like rhododendrons, and coffee and tea plants as well. In all cases, planting in recently cleared woodland exposes the plant to the fungus living on the tree roots left in the soil, which is what I had unknowingly done. The mushroom itself goes by the common name "honey mushroom," an edible mushroom enjoyed by many wild collectors, but that didn't make me feel any better.

Clearing the blueberry field.
BELOW Chet and Steve with the chipper.

For the next two years, I worked with the two plant pathologists that had identified the disease, trying to treat the plants, but nothing worked. I had to accept that the yield would never get high enough to break even, let alone make a profit. I either had to give up and quit blueberry farming, or remove all the plants, apply a fumigant to the soil, and replant. After a lot of thought, I chose the latter. In 1995, I closed the farm and began digging the plants out. By 1998, the process was complete. I decided to plant only half the field to save money. In 2001, I opened to the public with my brand new, one-and-a-half acres of blueberries free of oak root disease.

Foreground, blueberry field tree stumps; background, vegetable and strawberry field in rye grass.
BELOW Pulling stumps. | Digging holes for the blueberry plants.

For the past fifteen summers, patrons of all ages have enjoyed picking blueberries on our farm. It makes us happy to see all the fun they have. They frequently thank us for having the farm. Many families have made it a tradition to come and pick blueberries year after year. Some of the harvests have been large, and some not so large. Sometimes we know why that happens, and sometimes we don't. Despite our knowledge and experience, there are uncertainties and mysteries in every growing season. There are always challenges growing food. We now have three invasive insects here that have spread across the country after coming in from other parts of the world. They have no natural enemies and there are no non-chemical controls. They force us to respond with new measures to protect the crop.

The three-acre first blueberry planting.
BELOW The fumigation setup after the infected plants were removed.

The long-term future of the farm is unknown. I sold the farm to the town of Dennis in 2001, to prevent the land's development and protect the recharge area for a nearby town water well. I retained the right to farm as long as I wanted with exclusive use under a life estate. When I stop farming, the town may consider leasing the farm to someone else or establishing an additional community garden. It would be great if the land continued to produce food. Some projections of climate change say that by the end of the century, Cape Cod may have the same weather as South Carolina. That would not provide the necessary chilling requirement for our Northern highbush blueberry species to grow properly. A Southern highbush species could be planted and would do just fine. At the very least, if the farm does not continue in some way, the land will return to a forest, protecting the water supply as well as continuing to be a home for wildlife. I will be fine with that outcome also.

I have had the good fortune to be able to pursue my passion for farming and a life lived outdoors close to the land. I still enjoy doing conservation work for NRCS after thirty-three years. We all learn important life lessons in some way, no matter what we do. My teacher has been Nature. She has taught me patience, resilience, persistence, and creativity. It has not always been easy, nor did the project unfold as I imagined many years ago. Still, being able to cultivate such a beautiful plant, and provide its fruit to so many people, has made all the years of effort worthwhile.

—Steve

The present 1.5-acre second blueberry planting with wildflower meadow.

FLORA AND FAUNA OF *Hokum Rock Farm*

Worldwide, we are experiencing the extinction of wildlife and plant species, by some counts at a rate 1,000 times that of earlier eras. In our stewardship of the land, we respect the rights and well-being of the wildlife and plants species that share this habitat with us. On our eighteen-acre farmland and woods, we have identified many species over the years and will work to add to our biodiversity count in the years ahead; hopefully we will not have to subtract from it. Every season brings us the joy of discovery of the varied and vibrant biodiversity present here. Following is a partial list as of 2016:

Wild Flowers

Amaranth, Green - *Amaranthus retroflexus*
Arrow-Leaved Tearthumb (Native) -*Polygonum sagittatum*
Aster, Heath (Native) - *Symphyotrichum ericoides*
Aster, Late Purple (Native) - *Aster patens*
Aster, White Wood (Native) - *Eurybia divaricata*
Bearberry - *Arctostaphylos uva-ursi*
Black-Eyed Susan (Native) - *Rudbeckia hirta*
Blackberry (Native) - *Rubus fruticosus*
Blue/Old-Field Toadflax (Native) - *Linaria canadensis*
Broadleaf Dock - *Rumex obtusifolius*
Buttercup, Creeping - *Ranunculus repens*
Buttercup, Tall/ Meadow - *Ranunculus acris*
Butterfly-Weed (Native) - *Asclepias tuberose*
Canada Mayflower (Native) - *Maianthemum canadense*
Chicory-*Cichorium intybus*
Clematis (Native) - *Clematis virginiana*
Common Blue Violet (Native) - *Viola papilionacea*
Common Burdock - *Arctium minus*
Common Fleabane - *Erigeron philadelphicus*
Common Milkweed - *Asclepias syriaca*
Common Speedwell - *Veronica officinalis*
Common Sunflower - *Helianthus annuus*
Coreopsis/Long-Stalked Tickseed (Native) - *Coreopsis lanceolata*
Cypress, Spurge - *Euphorbia cyparissias*
Daisy Fleabane (Native) - *Erigeron annuus*
Daisy, Ox-Eye (Native) - *Leucanthemum vulgare*
Dandelion - *Taraxacum officinale*
Daylily - *Hemerocallis fulva*
Deptford, Pink - *Dianthus armeria*
Dewberry (Native) - *Rubus flagelleris*
Dwarf/Old-Field Cinquefoil (Native) -*Potentilla canadensis*
Everlasting Pea - *Lathyrus latifolius*
Field Bindweed - *Convolvulus arvensis*
Field Pansy (Native) - *Viola bicolor*
Field Thistle (Native) - *Cirsium discolor*
Forget-Me-Not, True - *Myosotis scorpiodes*
Foxglove, Common - *Digitalis purpurea*
Frostweed (Native) - *Helianthemum canadense*
Goldenrod, Early (Native) - *Solidago juncea*
Goldenrod, Seaside (Native) -*Solidago sempervirens*
Grape-hyacinth - *Muscari botryoides*
Heath Aster (Native) -*Symphyotrichum ericoides*
Hedge Bindweed (Native) -*Calystegia sepium*
Henbit - *Lamium amplexicaule*
Hop Clover - *Trifolium agrarium*
Horse-Nettle (Native) - *Solanum carolinense*
Indian-Pipe (Native) - *Monotropa uniflora*
Inkberry (Native) - *Ilex glabra*

Japanese Honeysuckle - *Lonicera japonica*
Lady's Slipper, Pink/Moccasin Flower
- *Cypripedium acaule*
Lady's Thumb - *Polygonum persicaria*
Lambsquarters - *Chenopodium album*
Late Purple Aster (Native)
-*Symphyotrichum patens*
Low Showy Aster (Native)
- *Eurybia spectabilis*
Lupine, Wild Blue (Native)
- *Lupinus perennis*
Milkweed, Blunt-Leaved (Native)
- *Asclepias amplexicaulis*
Mouse-Ear Hawkweed
- *Hieracium pilosella*
Mugwort - *Artemisia vulgarisa*
Mullein, Common - *Verbascum thapsus*
Mullein, Pink - *Lychnis coronaria*
Multiflora Rose - *Rosa multiflora*
Ox-Eye Daisy (Native)
- *Leucanthemum vulgare*
Periwinkle/Myrtle - *Vinca minor*
Plantain, English - *Plantago lariceolata*
Plantain-Leaved Pussytoes (Native)
- *Antennaria plantaginifolia*
Poison Ivy, Common (Native)
- *Toxicodendron radicans*
Pokeweed (Native) - *Phytolacca americana*
Primrose, Common Evening
- *Oenothera biennis*
Purple Dead-Nettle - *Lamium purpureum*
Purslane, Common - *Portulaca oleracea*
Queen Anne's lace - *Daucus carota*
Rabbit-Foot Clover - *Trifolium arvense*
Ragweed, Common
- *Ambrosia artemisiifolia*
Red Clover - *Trifolium pratense*
Red Sorrell - *Rumex acetosells*
Rock Cress, Lyre-leaved-*Arabis lyrata*
Rocky Mountain Bee Flower
- *Cleome serrulata* (once)
Scotch broom - *Cytisus scoparius*

Shepard's Purse - *Capsella bursa-pastoris*
Smartweed, Long-Bristled
- *Polygonum caepitosum*
Spiderwort - *Tradescantia virginiana*
Spotted knapweed - *Centaurea maculos*
Spotted Wintergreen (Native)
- *Chimaphila maculate*
St. John's Wort, Common
- *Hypericum perforatum*
Star-of-Bethlehem
- *Ornithogalum umbellatum*
Starflower (Native) - *Trientalis borealis*
Toadflax (Butter-and-Eggs)
- *Linaria vulgaris*
Tulip - *Tulipa gesneriana*
Virginia Dayflower
- *Commelina virginica*
Virginia Creeper (Woodbine)
- *Parthenocissus quinquefolia*
Virginia Spiderwort (Native)
- *Tradescantia virginiana*
Vetch, Cow - *Vicia cracca*
Vetch, Purple - *Vicia americana*
White Campion - *Silene latifolia*
White Clover - *Trifolium repens*
White Wood Aster (Native)
- *Eurybia divaricata*
Wood Anemone (Native)
- *Anemone quinquefolia*
Yarrow (Native) - *Achillea millefolium*
Yellow Stargrass/Common Goldstar
(Native) - *Hypoxis hirsute*
Yucca - *Yucca filamentosa*

Grasses, Ferns & Sedges

Bristly Foxtail - *Setaria verticillata*
Brome (Native) - *Bromus ciliatus*
Broom Sedge (Native)
- *Andropogon virginicus*
Canada Bluegrass - *Poa compressa*
Deer-Tongue (Native)
- *Dichanthelium clandestinum*
Field Garlic - *Allium vineale*
Lady Fern - *Athyrium filix*
Little Barley - *Hordeum pusillum*
Little Bluestem - *Andropogon scoparius* *(aka/Schizachyrium scoparium)*
Meadow Fescue - *Festuca elatior*
Oats - *Avena sativa*
Old Witch Grass (Native)
- *Panicum capillare*
Panic Grass (Native)
- *Panicum lanuginosum*
Poverty Grass (Native)
- *Aristida dichotoma*
Quack Grass (Native) - *Agropyron repens*
Redtop (Native) - *Agrostis alba*
Sedge (Native) - *Carex scoparia*
Sensitive Fern - *Onoclea sensibilis*
Sweet Fern - *Comptonia peregrina*
Sweet Vernal Grass
- *Anthoxanthum odoratum*
Switch Grass - *Panicum virgatum*
Timothy - *Phleum pratense*
Wild Rye - *Elymus virginicus*
Yellow Nutsedge - *Cyperus esculentus*

Other

Mosses, Various, including:
Club Moss - *Lycopodium obscurum*
Cushion Moss - *Selaginella kraussiana*
Mushrooms, Various, including:
Giant Puffball Mushroom
- *Calvatia gigantea*
Little Toadstool Mushroom
(several varieties) - *Amanita muscaria*
Possibly:
Cinnamon Webcap Mushroom
- *Cortinarius cinnamomeus*
False Parasol Mushroom
- *Chlorophyllum molybdites*
Heath Waxcap Mushroom
- *Hygrophorus laetus*
King Bolete - *Boletus edulis*
Parasol Mushroom -*Macrolepiota procera*
Sickener Mushroom - *Russula emetica*
Lichen, Various, including:
Lichen, Common Green Shield
- *Flavoparmelia caperata*
Lichen, Dixie Reindeer Shield
- *Cladina subtenuis*
Lichen, Old Man's Beard - *Usnea strigosa*
Lichen, Peppered Rock Shield
- *Xanthoparmelia conspersa*
Lichen, Texas Shield
- *Canoparmelia texana*

Trees & Shrubs

American Holly - *Ilex opaca*
Arborvitae, Eastern White Cedar
- *Thuja occidentalis*
Ash - *Fraxinus omus*
Atlantic White Cedar
- *Chamaecyparis thyoides*
Autumn Olive - *Elaeagnus umbellata*
Black Cherry - *Prunus serotina*
Black Oak, Eastern - *Quercus velutina*
Blueberries, Common Lowbush (Native)
- *Vaccinium angustifolium*
Blueberries, Highbush (wild & cultivated), *Vaccinium corymbosum*
Cherry, Pin - *Prunus pensylvanica*
Common Chokecherry
- *Prunus virginiana*
Common Juniper - *Juniperus communis*
Eastern Hemlock - *Tsuga canadensis*
Eastern Red Cedar - *Juniperus virginiana*
Eastern White Pine - *Pinus strobus*
Elm, American - *Ulmus Americana*
Elm, Slippery - *Ulmus rubra*
Forsythia - *Forsythia x intermedia*
Honeysuckle - *Lonceria japonica*
Lilac - *Syringa vulgaris*
Locust, Black - *Robinia pseudoacacia*
Locust, Honey - *Gleditsia triacanthos*
Maple, Norway - *Acer platanoides*
Oak, Northern Red - *Quercus rubra*
Oak, White - *Quercus alba*
Pine, Pitch - *Pinus rigada*
Pine, Red - *Pinus resinosa*
Sassafras - *Sassafras albidum*
Scotch Pine - *Pinus sylvestris*
Sugar Maple - *Acer saccharum*
Sumac, Staghorn - *Rhus typhina*
Sweet Pepper-Bush (Native)
- *Clethra alnifolia*
Sycamore, American
- *Platanus occidentalis*

Insects & Arthropods

Ants (various) - *Formicidae*
Aphids - *Aphidoidea*
Banded Woollybear Caterpillar Moth
- *Pyrrharctia isabella*
Bee, American Bumble
- *Bombus pennsylvanicus*
Bee, Common Eastern Bumble
- *Bombus impatiens*
Bee, Honey - *Apis mellifera*
Butterfly, Baltimore checkerspot
- *Euphydryas phaeton*
Butterfly, Cabbage White - *Pieris rapae*
Butterfly, Monarch Butterfly
- *Danaus plexippus*)
Butterfly, Mourning cloak (Camberwell beauty/Brit.) - *Nymphalis antiopa*
Butterfly, Spring azure - *Celastrina ladon*,
Cicada, Dog Day - *Tibicen canicularis*

Click Beetles - *Elateridae*
Clover mites - *Bryobia praetiosa*
Cricket, Field - *Gryllus pennsylvanicus*
Cricket, House - *Acheta domestica*
Cricket, Spotted Camel
- *Ceuthophilus maculatus*
Damselfly, Familiar bluet (Endangered)
- *Enallagmam civile*
Damselfly, Green darner - *Anax junius*
Dragonfly, Blue-eyed darner,
- *Aeshna multicolor*
Dragonfly, Common Whitetail
- *Plathemis(Libellula)*
Earthworm - *Lumbricina*
Earwig, Common - *Forficula auricularia*
Firefly, Common Eastern
- *Photinus pyralis*
Flies - *Diptera*
Fly, Scorpion - *Mecoptera*
Fly, Spotted Wing Drosophilia Fruit
- *Drosophila suzukii*
Grasshopper, Carolina locust
- *Dissosteira carolina*
Green Lacewings - *Chrysopa perta*
Hornet, Bald-Faced
- *Dolichovespula maculata*
Hover (Flower) flies - *Syrphidae*
Katydid - *Microcentrum rhombifolium*
Japanese Beetle - *Popillia japonica*
Ladybug, Asian - *Harmonia axyridis*
Ladybug, Seven-spotted Lady Beetle
- *Coccinellidae septempunctata*
Ladybug, Two-spotted Lady Beetle
- *Adalia bipunctata*
Leafhopper - *Cicadellidae*
Live-oak Root Borer
- *Archodontes melanopus*
Long Horn Beetle/Red Milkweed Borer
- *Tetraopes tetrophanus*
May Beetle (June Bug) - *Phyllophaga*
Meadow Spittlebug (Froghopper)
- *Philaenus spumarius*
Mosquitos - *Culicidae*
Moth, Milkweed Tussock (Tiger)
- *Euchaetes egle*
Moth, Orange Sulphur - *Colias eurytheme*
Moth, Tiger - *Arctiinae*
Moth, Winter - *Operophtera brumata*
Moth, Underwing - *Catocala*
Robber Fly - *Asilidae laphria*
Sowbug - *Oniscidea*
Spiders (various) - *Araneae*
Springtails - *Collembola*
Stink Bug, Brown Marmorated
- *Halyomorpha halys*
Ticks - *Ixodidae*
Tiger Beetle, Six-Spotted
- *Cicindela sexguttata*
Wasp, Paper - *Polistes exclamens*
Wasp, Yellow Jacket - *Vespula maculifrons*

Reptiles

American Toad - *Anaxyrus americanus*
Eastern Box Turtle (Endangered)
- *Terrapene carolina carolina*
Salamander, Eastern Red-backed
- *Plethodon cinereus*
Snake, Black Racer - *Coluber constrictor*
Snake, Eastern Garter
- *Thamnophis sirtalis*
Snake, Eastern Milk
- *Lampropeltis triangulum*

Birds

American Crow - *Corvus brachyrhynchos*
American Robin - *Turdus migratorius*
Baltimore Oriole - *Icterus galbula*
Bluebird, Eastern - *Sialia sialis*
Bluejay - *Cyanocitta cristata*
Broad-winged Hawk - *Buteo platypterus*
Brown-headed Cowbird - *Molothrus ater*
Canada Goose - *Branta canadensis*
Cardinal - *Cardinalidae*
Catbird, Gray - *Dumetella carolinensis*
Chickadee, Black-capped
- *Poecile atricapillus*
Cooper's Hawk - *Accipiter cooperii*
Finch, House - *Haemorhous mexicanus*
Finch, Purple - *Haemorhous purpureus*
Goldfinch, American - *Spinus tristis*
Grackle, Common - *Quiscalus quiscula*
Great Blue Heron - *Ardea herodias*
Gull, Herring, (daily flyover)
- *Larus smithsonianus*
Mockingbird, Northern
- *Mimus polyglottos*
Mourning Dove - *Zenaida macroura*
Northern Harrier (Marsh Hawk)
- *Circus cyaneus*
Nuthatch, Red-breasted - *Sitta canadensis*
Osprey (daily summer flyover)
- *Pandion haliaetus*
Owl, Eastern Screech - *Megascops asio*
Owl, Great Horned - *Bubo virginianus*
Red-tailed Hawk - *Buteo jamaicensis*
Red-winged Blackbird
- *Agelaius phoeniceus*
Ruby-throated Hummingbird
- *Archilochus colubris*
Screech Owl, Eastern - *Megascops asio*
Sparrow, American Tree - *Spizella arborea*
Sparrow, Chipping - *Spizella passerina*
Sparrow, House - *Passer domesticus*
Sparrow, Song - *Melospiza melodia*
Starling, Common - *Sturnus vulgaris*
Swallow, Barn - *Hirundo rustica*
Tern, Common (daily summer flyover)
- *Sterna hirundo*
Tufted Titmouse - *Baeolophus bicolor*
Turkey - *Meleagris gallopavo*
Turkey Vulture (flyovers) - *Cathartes aura*
Warbler, Common yellowthroat
- *Geothlypis trichas*
Woodpecker, Downey - *Picoides pubescens*
Woodpecker, Northern Flicker
- *Colaptes auratus*
Wren, Carolina-*Thryothorus ludovicianus*

Fauna

American Black Bear (possible)
- *Ursus americanus*

Bat, Little Brown, *Myotis lucifugus*
Cat - *Felis catus*
Chipmunk - *Tamias minimus*
Coyote/Coywolf/Eastern Coyote
- *Canis latrans x Canis lupus*

Deer, White-tail *Odocoileus virginianus*
Dog - *Canis familiaris*
Fox, Red - *Vulpes vulpes*
Groundhog - *Marmota monax*
Human - *Homo sapien*
Mole, Eastern - *Scalopus aquaticus*
Mouse, House - *Mus musculus*
Opossum - *Didelphimorphia*
Rabbit, Cottontail - *Sylvilagus*
Raccoon - *Procyon lotor*
Skunk - *Mephitidae*
Squirrel, Eastern Grey
- *Sciurus carolinensis*

Vole, Meadow - *Microtus pennsylvanicus*

BLUEBERRY FACTS

A half-cup of blueberries has only forty-two calories, 10.7 grams of carbohydrates, 7.4 grams of sugar, 1.8 grams of dietary fiber, and 0.6 grams of protein, as well as manganese, iron, and minerals. They are fat free.

Blueberries are ranked the #1 fruit for antioxidant benefits. They are an excellent source of vitamin C, which helps maintain a healthy immune system

Anthocyanin is the plant chemical in the leaves, stems, roots, flowers, and berries that is responsible for the berries' blue pigmentation and the red leaves and canes in autumn. Recent studies show its benefits in cardiovascular health and may help deter cancer-causing free radicals. Anthocyanins can help ward off urinary tract infections in the same way cranberries do, inhibiting bacteria from binding to cell tissues. Recent studies are showing that this same chemical can improve brain health and reduce memory decline.

Blueberries contain a high level of pectin, a natural thickening agent, which causes them to gel when cooked without the aid of commercial products. The average content is 0.3 g per half-cup.

Blueberries are an important source of nutrition and hydration not only for humans, but for birds, insects, and various mammals, including deer, foxes, coyotes, squirrels, and reptiles, like turtles, frogs, and salamanders.

Only three fruits are native to North America—blueberries, cranberries, and Concord grapes. Botanists say that blueberries have been around for 13,000 years. Native Americans gave blueberries to the early colonists in New England to see them through their first winter.

In 1999, the USDA declared July National Blueberry Month.

A single blueberry bush can produce 6,000 berries per year. At our farm, we grow the following varieties of highbush blueberries: Duke, Spartan, Coville, Blue Crop, Blue Gold, and Late Blues. We also recommend Nelsons.

See Cornell University's *Blueberry Variety Review* for more information at: www.fruit.cornell.edu.

BLUEBERRY TIPS

PLANTING

- plant 3–6 feet apart/8–12 feet between rows
- plant at least two different varieties for extended harvest
- shrubs are mature after eight years and can live for decades with proper care

SUN

- blueberry plants require eight hours of daily sun in an open location

SOIL

- acidic soil, moist, with good drainage and high organic matter
- add sulfur to pH 4.5–5.5 if necessary

WATER

- 1 inch per week in growing season
- 2 inches during ripening
- drip irrigation or soaker hose is best

WEED AND FERTILIZE

- weed continually
- fertilize at bloom and again one month later—not in late season

MULCH

- leaf mulch is okay, 2 to 4 inches deep, all year, especially November and December

PRUNING

- New plants: light spring pruning for three years; pick berries on fourth year
- Established plants: prune twenty percent of shrub annually every March and April
- first prune oldest large branches with few set buds—these are unproductive
- then prune weakest and most congested of new canes, keeping new ground shoots
- prune small, damaged twigs
- pruning should invigorate growth, allow sunlight in, and open plant to circulating air

NETTING

- net shrubs when berries start to color

BERRIES

- wait six days after berries achieve full blue color before picking for sweeter taste
- average yield—eight pints per plant
- blueberries should not be washed until ready to eat, as berries tend to mold when moist without proper air circulation
- berries can sit at room temperature for two days, be refrigerated for one week, and frozen for one year or more (but will be pulpy)
- berries washed before freezing should be air-dried; they can be washed after freezing, but this can be messy (dusting with flour will keep moisture contained for use in recipes)

HOKUM ROCK FARM FAVORITE BLUEBERRY RECIPES

A baker's dozen . . .

Apple Blueberry Crisp

½ cup butter, melted and divided
½ cup brown sugar, packed and divided
1 tablespoon lemon juice
1 tablespoon flour
¾ teaspoon cinnamon, divided
4 cups tart apples, pared, cored, and sliced
2 cups blueberries
1 cup quick oats
¼ cup chopped walnuts
¼ teaspoon ginger
whipped cream or ice cream

Preheat oven to 375 degrees.

In a large bowl combine ¼ cup melted butter, ¼ cup brown sugar, lemon juice, flour, and ½ teaspoon cinnamon. Add apples and washed blueberries, tossing to coat. Place in 8-inch square baking pan. In a small bowl combine oats, ¼ teaspoon melted butter, ¼ cup brown sugar, walnuts, ¼ teaspoon cinnamon, and ginger until well-blended. Sprinkle over fruit. Bake until fruit is tender and topping is crisp and brown. Serve warm with whipped cream or ice cream. Yields 6 servings.

Big Blueberry Muffins

1½ cups all-purpose flour
¾ cup white sugar
1/2 teaspoon salt
2 teaspoons baking powder
1/3 cup vegetable oil
1 egg
1/3 cup milk
1 cup fresh blueberries

Crumb topping:
1/2 cup white sugar
1/3 cup all-purpose flour
¼ cup butter, cubed
1½ teaspoons ground cinnamon

Preheat oven to 400 degrees. Grease muffin cups or line with muffin liners.

Combine flour, sugar, salt, and baking powder. Mix together oil, egg, and milk, and combine with flour mixture. Fold in blueberries. Fill muffin cups right to the top (skip some of the cups if necessary).

Mix crumb topping ingredients with a fork and sprinkle over the muffins. Bake for 20 to 25 minutes.

Blueberry Banana Oatmeal Bread

2 eggs, beaten
⅓ cup vegetable oil
1 cup ripe, mashed bananas
1½ cups flour
⅔ cup sugar
2¼ teaspoons baking powder
1 teaspoon wheat germ
1/2 teaspoon salt
1/2 cup quick oats
1 cup blueberries

Preheat oven to 350 degrees.

Combine eggs, oil, and bananas. Sift together flour, sugar, baking powder, wheat germ, and salt. Mix in oats. Add dry ingredients to banana mixture, stirring until just combined. Gently fold in washed berries. Pour into greased and floured 9 × 5-inch loaf pan. Bake for 1 hour. Cool completely before cutting; best if baked a day ahead.

Blueberry Cornmeal Pancakes

1 cup all-purpose flour
½ cup cornmeal
2 tablespoons sugar
1¼ teaspoons baking powder
1/2 teaspoon baking soda
¼ teaspoon fine sea salt
1¼ cups buttermilk
4 tablespoons unsalted butter, melted, plus more for the skillet
1 egg
1 pint blueberries

Whisk together the flour, cornmeal, sugar, baking powder, baking soda, and salt in a large bowl. Whisk together the buttermilk, melted butter, and egg in a medium bowl. Whisk the wet ingredients into the dry ingredients, then fold in the blueberries.

Heat a large nonstick skillet or griddle over medium heat. Dip a paper towel in melted butter and wipe it on the bottom of the skillet to lightly grease. For each pancake, spoon a scant ½ cup of batter onto the prepared skillet, frying two to three minutes per side. Repeat, buttering the skillet between batches. Keep stack warm in oven. Serve with butter and maple syrup.

Blueberry Cranberry Cobbler

2 cups cranberries
2 cups blueberries
1 tablespoon fresh lemon juice
2 tablespoons cassis liquor
1 cup flour
1 cup sugar
¼ teaspoon salt
1 large egg
6 tablespoons melted butter

Preheat oven to 400 degrees.
Grease a 10 × 6 pan.

Combine cranberries, blueberries, lemon juice, and cassis and spread in pan. Stir together flour, sugar and salt. Add lightly beaten egg and mix until coarse, then spread over berry mixture. Drizzle with melted butter and bake for 30 minutes or until light brown and bubbling. Serve with cream, whipped cream, or ice cream.

—Courtesy of patron Susan D. Arayas, adapted from Bilner's Cookbook

Blueberry Crepes

Crepes:
1 egg
2 tablespoons flour
3 tablespoons low-fat milk

Beat ingredients together and set aside. Grease crepe skillet or omelet pan and warm over medium heat. Pour in 2 tablespoons batter and quickly rotate pan to cover bottom evenly. Cook about 30 seconds until surface looks dry. Turn carefully and cook second side. Turn onto plate and keep warm while making more crepes.

Filling:
¼ cup sour cream
¼ cup plain or vanilla yogurt
1 cup blueberries

Combine sour cream and yogurt. Wash berries and set aside. Spoon 3–4 tablespoons berries into center of each crepe and add yogurt mixture. Roll crepes and arrange on plate. 4 servings.

Blueberry Pie

Pastry:
2 cups flour
½ teaspoon salt
¾ cup shortening
6–7 tablespoons ice water

Blueberry filling:
¾ cup white sugar
¾ cup packed brown sugar
¼ cup flour
3 tablespoons quick-cooking tapioca
¼ teaspoon ground allspice
¼ teaspoon ground cloves
4 cups washed blueberries
3 tablespoons blueberry (or other) liqueur

To make pastry:
In large bowl combine flour and salt. With pastry blender or two knives, cut in shortening until lumps are the size of peas. Add ice water, a tablespoon at a time, tossing with a fork until all is moist. Gather dough into ball. Divide into ⅓ and ⅔ portions. Flatten both pieces into disc shapes. (Do not over-handle dough, as it will dry out.) Wrap and refrigerate while making filling.

To make blueberry filling:
In medium bowl combine white sugar, brown sugar, 1 cup flour, tapioca, allspice, and cloves. Mix until blended. Pour over blueberries in large bowl. Mix gently. Drizzle with liqueur, then fold gently.

To assemble:
Preheat oven to 375 degrees.

On surface lightly dusted with flour, roll the ⅔ pastry disc. Gently fold disc in half and lay out in a 9-inch pie pan, unfolding to fill pan. Spoon blueberry filling onto pastry. Roll out ⅓ pastry, fold, lay over berry mixture, and unfold to cover. Trim and flute edges—not too thin or it will crumble when baked. Make slits for steam in crust. Bake for 55–65 minutes until bubbling in center and crust is browned. Yields 6 servings.

Daryl's Lemon Pound Bundt Cake with Blueberries

1 cup blueberries
1 tablespoon flour for tossing
2¼ cups flour
(½ whole wheat works well)
1½ cups sugar
½ teaspoon salt
½ teaspoon baking soda
1 teaspoon vanilla
1 cup butter, softened
8 ounces sour cream
(or plain yogurt)
3 large eggs at room temp.
grated lemon rind from 1 lemon

Lemon glaze:
1 cup confectioners sugar
2 tablespoons lemon juice

Preheat oven to 325 degrees.

Toss berries with 1 tablespoon flour and set aside. Place remaining ingredients in a mixing bowl. Blend at low speed, then beat at medium speed. Fold in berries carefully. Pour batter into greased and floured Bundt pan (or large loaf pan). Bake for 60–70 minutes.

Let cake cool for at least 15 minutes. Combine lemon glaze ingredients and drizzle over cake.

—Courtesy of Daryl Bladen

Jeri's Blueberry Jam

4½ cups Steve's wonderful
blueberries, washed
2 tablespoons fresh lemon juice
7 cups raw sugar
6 ounces Certo liquid pectin
(2 pouches)

Combine all ingredients in saucepan. Cook until sugar dissolves and mixture begins to boil. Cook at rolling boil for 2 minutes, stirring continuously.

Remove from heat. Let stand for 5 minutes. Skim off foam.

Pour into sterilized jars, leaving ¼-inch at top. Cover tightly with lids. Makes approximately 5 cups.

—Courtesy of patron Jeri Lavender. Jeri advises following the pectin instructions that come with each package. Jam takes practice to perfect.

John's Flax Seed Blueberry Muffins

1¾ cups whole wheat pastry flour
¾ teaspoon salt
⅓ cup sugar
2 teaspoons baking powder
2 eggs
¼ cup melted butter
¾ cup milk
1 cup blueberries
1 teaspoon grated orange
  or lemon rind
¼–½ cup flax seeds

Preheat oven to 400 degrees.

Sift dry ingredients together. In separate bowl beat eggs, then add butter and milk and beat together. Combine liquid and dry ingredients. Gently fold in washed blueberries. Add citrus rind and flax seeds. Fill well-greased muffin cups two-thirds full. Bake for 20–25 min. Yields 2 dozen.

O'Brien's Easy Blueberry Breakfast Cake

2 cups flour
3 teaspoons baking powder
1 teaspoon salt
2½ tablespoons butter
¾ cup sugar
1 egg plus milk to make 1 cup
1 cup blueberries

Preheat oven to 400 degrees.

Sift flour, baking powder, and salt into bowl. Cut in butter and sugar thoroughly. Beat in egg/milk mixture just enough to wet flour. Mix in berries. Spread evenly in 8-inch square pan or loaf pan. Sprinkle with sugar. Bake for 35–40 minutes.

—Courtesy of Jack & Judy

Henrietta's Easy Graham Crust Blueberry Pie

2 tablespoons butter
2 tablespoons flour (heaping)
1 cup sugar
1 quart blueberries
1 9-inch store-bought graham cracker pie crust

Melt butter in pan. Mix flour and sugar together and add to half of the berries (2 cups). Cook the berries in the melted butter until thickened. Cool. Put remaining berries in the pie shell and pour the cooked berry mixture over them. Chill. Serve with whipped cream.

—Courtesy of patron Lesley Kramer

Hokum Rock Farm's Blueberry Buckle

EVERYONE'S FAVORITE

¾ cup sugar
¼ cup shortening
½ cup milk
2 cups flour
2 teaspoons baking powder
½ teaspoon salt
2 cups blueberries
1 egg

Crumb topping:
½ cup sugar
⅓ cup flour
½ teaspoon cinnamon
¼ cup cold cubed butter

Preheat oven to 375 degrees.

Grease and lightly flour 9-inch square pan. Wash and drain berries and set aside. In large bowl mix sugar, shortening (4 tablespoon soft butter can be substituted), and egg thoroughly. Stir in milk. In small bowl, sift flour, baking powder, and salt. Mix into large bowl with other ingredients. Blend in berries carefully. Spread mixture into prepared pan.

To prepare topping, cube butter, then mix in other ingredients. To achieve desired consistency and taste, add a touch more sugar or flour while mixing. Dab, sprinkle, or spread (depending on consistency) over berry mixture. Bake for 45–50 minutes until toothpick inserted in center comes out clean. When cooled, cut into 3-inch squares. One patron affectionately named this the "little black dress" recipe.

BIBLIOGRAPHY & RESOURCES

Beston, Henry. *The Outermost House.* New York: Henry Holt and Company, 1928.

_____. *Northern Farm: A Chronicle of Maine.* New York: Rinehart & Co., 1948.

Brockman, C. Frank. *Trees of North America.* New York: Golden Press, 1968.

Brown, Lauren. *Grasses: An Identification Guide.* Boston: Houghton Mifflin Company, 1969.

Campisi, Jack. *The Mashpee Indians, Tribe on Trial.* New York: Syracuse University Press, 1991.

Carter, Kate. *Wildflowers of Cape Cod.* Woodstock,VT: The Countryman Press, 2008.

Coville, Frederick. "Experiments in Blueberry Culture." Bulletin No. 193. Washington, DC: USGPO, 1910.

Cronon, William. *Changes in the Land, Indians, Colonists, and the Ecology of New England.* New York: Farrar, Straus and Giroux, 2003.

Cumbler, John T. *Cape Cod, An Environmental History of a Fragile Ecosystem.* Amherst: UMass Press, 2014.

Eck, Paul, Editor, Norman F. Childers, Co-editor. *Blueberry Culture.* New Brunswick, NJ: Rutgers University Press, 1966.

_____. *Blueberry Science.* New Brunswick, NJ: Rutgers University Press, 1988.

Elias, Thomas S. *The Complete Trees of North America: Field Guide and Natural History.* New York: Van Nostrand Reinhold, 1980.

Gough, Robert E. *The Highbush Blueberry and its Management.* New York: The Haworth Press, Inc., 1994.

Hay, John. *Nature's Year: The Seasons of Cape Cod.* New York: Doubleday & Co., Inc., 1961.

_____. *The Run.* New York: Doubleday & Co., Inc., 1959.

Heinrich, Bernd. *Winter World: The Ingenuity of Animal Survival.* New York: Harper Collins Publishers, Inc., 2003.

_____. *Bumblebee Economics.* Cambridge: Harvard University Press, 1979.

_____. *Summer World.* New York: Harper Collins Publishers, 2009.

Hill, Lewis, Leonard Perry. *The Fruit Gardener's Bible.* North Adams, MA: Storey Publishing, 2011.

Hinds, Harold R., Wilfred A. Hathaway. *Wildflowers of Cape Cod.* Chatham, MA: The Chatham Press, Inc.,1968.

Isenberg, Nancy. *White Trash: The 400-Year-Old Untold History of Class in America.* New York: Viking, 2016.

Kittredge, Henry C. *Shipmasters of Cape Cod.* Boston: Houghton Mifflin Co., 1935.

Lewis, Sara. *Silent Sparks: The Wondrous World of Fireflies.* Princeton, NJ: Princeton University Press, 2016.

Mandell, Daniel R. *King Philipp's War: Colonial Expansion, Native Resistance, and the End of Indian Sovereignty*. Baltimore: John Hopkins University Press, 2010.

McPhee, John. *The Pine Barrens*. New York: Farrar, Straus, and Giroux, 1968.

Merchant, Carolyn. *Ecological Revolutions, Nature, Gender, and Science in New England*. Chapel Hill: The University of North Carolina Press, 1989.

Miller, Orson K. Jr. *Mushrooms of North America*. New York: E. P. Dutton, 1979.

Minick, Jim. *The Blueberry Years: A Memoir of Farm and Family*. New York: St. Martin's Press, 2010.

Oberhauser, Karen S., Kelly R. Nail, Sonia Altizer, Editors. *Monarchs in a Changing World: Biology and Conservation of an Iconic Butterfly*. New York: Cornell University Press, 2015.

Peterson, Roger Tory. *Wildflowers: Northeastern/ North-central North America*. Boston: Houghton Mifflin Company, 1968.

_____. *Peterson Field Guide to Birds of North America*. Boston, Massachusetts: Houghton Mifflin Co., 2008.

Pritts, Marvin P., James F. Hancock, Editors. *Highbush Blueberry Production Guide*. New York: Northeeast Regional Agricultural Engineering Service NRAES-55, 1992.

Rovelli, Carlo, trans. Simon Carnell and Erica Segre. *Seven Brief Lessons on Physics*. New York: Riverhead Books, 2016.

Russell, Howard S. *A Long, Deep Furrow, Three Centuries of Farming in New England*. Hanover, NH: University Press of New England. 1976.

Tallamy, Douglas W. *Bringing Nature Home: How You Can Sustain Wildlife With Native Plants*. Portland, OR: Timber Press, 2007.

Thoreau, Henry David. *Wild Fruits*, edited by Bradley P. Dean. New York: Norton, 2000.

_____. *Cape Cod*. New York: Penguin Random House, 1865/1987.

Tiner, Ralph W. *A Field Guide to Coastal Wetlands Plants of the Northeastern United States*. Amherst, MA: University of Massachusetts Press, 1987.

Trehane, Jennifer. *Blueberries, Cranberries and Other Vacciniums*. Portland, OR: Royal Horticultural Society, 2004.

Trull, Peter. *Coyotes in the Neighborhood: An Informational Guide to the Habits and Life History of the Eastern Coyote*. Cape Cod, MA: Shank Painter Publishing, 2002.

_____. *A Guide to the Common Birds of Cape Cod*. Cape Cod, MA: Shank Painter Publishing, 1994.

Schwarzman, Beth. *The Nature of Cape Cod*. Hanover, NH: University Press of New England, 2002.

Uva, Richard H., Joseph C. Neal, Joseph M. DiTomaso. *Weeds of the Northeast*. New York: Cornell University Press, 1997.

Wilson, E. O. *Half Earth, Our Planet's Fight for Life*. New York: W. W. Norton & Co., 2016.

_____. Editor. *Biodiversity*. Washington: National Academy Press, 1988.

POETRY CITED

Albright, Daniel, Ed. *W. B.Yeats:The Poems*, "Coole Park, 1929." London: J. M. Dent & Sons Ltd., 1996.

______. Ed. *e. e. cummings, Poems 1924–1954*, First Complete Edition. New York: Harcourt Brace, 1968.

Blodgett, Harold W., and E. Sculley Bradley, Editors. *Walt Whitman, Leaves of Grass*, "Song of Myself." New York: W. W. Norton & Company, 1965.

Doty, Mark. *Deep Lane*. "Deep Lane iv." New York: W. W. Norton & Company, 2015.

Erdman, David V., Editor. *The Complete Poetry and Prose of William Blake*, "Auguries of Innocence." New York: Anchor Press, 1988.

Heaney, Seamus. *The Essential Wordsworth*. "Ode: Intimations of Immortality" from *Recollections of Early Childhood*. New York: Ecco Press, 1988.

Johnson, Thomas H., Editor. *The Complete Poems of Emily Dickinson*, "It's All I Have To Bring Today # 26." Boston: Little, Brown & Co., 1961.

Lathem, Edward Connery, Editor. *The Poetry of Robert Frost*, "Dust of Snow," "Reluctance," "Nothing Gold Can Stay." New York: Holt, Rinehart and Winston, 1969.

Milford, Nancy, Editor. *The Selected Poetry of Edna St. Vincent Millay*, "Spring." New York: Modern Library, 2001.

Rilke, Rainer Maria. Trans. Stephen Mitchell. *The Selected Poetry of Rainer Maria Rilke*. New York: Vintage Books, 1984.

Seferis, George, trans. Rex Warner. *Poems*. Boston: Little, Brown and Co., 1964.

Thomas, Dylan. *A Child's Christmas in Wales*. New York: New Directions Publishing, 1982.

ONLINE RESOURCES:

American farming statistics are compiled by census through the United States Department of Agriculture (USDA) every five years: www.agcensus.usda.gov.

Association for the Preservation of Cape Cod's publication, "Agricultural Land Use on Caoe Cod: Looking to the Future," Katherine Beauchamp and Maggie Geist, 2011: www.apcc.org.

Astronomy: www.curious.astro.cornell.edu.

Barnstable County (Cape Cod) history, "The History of Barnstable County, 1620–1890," Simeon L. Deyo: (full text) www.archive.org.

Bird Science: www.birds.cornell.edu.

Blueberry Varieties Review: Cornell University College of Agriculture and Life Sciences/ Cornell Cooperative Extension: www.fruit.cornell.edu.

Bumblebees and honeybees—their differences: http://bumblebeeconservation.org.

Buy Fresh, Buy Local, a service of the Barnstable County Extension Office: www.BuyFreshBuyLocalCapeCod.org.

Cape Cod Extension Service, home of our entomologist Larry Dapsis: www.capecodextension.org.

Cape & Islands Farm Bureau, member of the Massachusetts Farm Bureau Federation: www.capecodfarms.com.

Colony collapse disorder in honeybees: ww.ars.usda.gov.

Cranberries: Our friends Leo, Evangeline, and Andrea Cakounes's Harwich farm and cranberry bog: www.cranberrybogtours.com; our friend Annie Walker's Dennis farm: www.anniescrannies.com.

Farm to Schools program through local agricultural efforts at sustainability and education: www.sustainablecape.org.

Historical Society of Old Yarmouth: www.hsoy.org.

History of blueberry cultivation by Rutgers University: www.njfarmfresh.rutgers.edu.

Insect identification for the casual observer: www.insectidentification.org.

Monarch butterflies and milkweed: www.ed.fnal.gov.

Old Farmer's Almanac, resource for farming information and folklore, including names of monthly full moons, and science of January thaw: www.farmersalmanac.com.

Southeastern Massachusetts Agricultural Partnership: www.semaponline.org

Spotted wing drosophilia and invasive fruit fly information: www.extension.umass.edu.

US Highbush Blueberry Council: www.blueberrycouncil.org

Whitesbog Preservation Trust, home of Elizabeth White's family bog where blueberries were first cultivated with Frederick Coville: www.whitesbog.org.

Winter moth facts: Center for Agriculture, Food and the Environment: www.ag.umass.edu.

Xerces Society Project Milkweed for nationwide effort to restore milkweed: www.xerces.org/milkweed.

Our blueberry farm website: www.hokumrockfarm.com.

Acknowledgments

Special thanks are owed to many people who have supported our farm through the years, especially our families, friends, staff, and patrons. Our gratitude is boundless. An additional note of thanks to:

Fred Barnicoat
Frank Caruso
Ed and Rachel (Carey) Harper
Barbara Khouri
George Kourian
Harry MacDonald
Rhoda, Tanney, & Howard Marks
Joan, Kevin, and Sean Micka
Dave Simser
Our Tibetan friends, Chewan, Tsering, Karma, and Nadoop
The Cape Wildlife Center
Whitesbog Preservation Trust
The Historical Societies of the towns of Brewster, Dennis, and Yarmouth
The United States Department of Agriculture
The Cape & Islands Farm Bureau
The Barnstable County Extension Service
The National Agricultural Library, Rare and Special Collections
All the libraries of Cape Cod, especially Brewster Ladies' Library, Eastham Library, and Jacob Sears Memorial Library

In supporting our efforts with this book, we would like to acknowledge the help and encouragement of Larry Dapsis, Peter Trull, John Tunney; our insightful and kind editors Catherine Mallette, Jesse Marth, Cheryl Weber; and especially Peter Schiffer, for saying yes.

Our Loyal Farm Helpers

Left to Right
Sandra
Bob
Mickayla
Stella
Bud
Lynn
Ray
Martha
David
Audrey
Chewan & Tsering
Leo
Owen, Liz, Barb & Deb
John
Gregg
Chip
Cy
Steve, Linda, Natalie, and Chloe
Anthony
Dave
Mary Jane
Peter
Evangeline

Red Sox

Janice Riley is a writer, photographer, and web master. She worked as a journalist for more than thirty years, starting out with a small newspaper. She then became a documentary editor, director, videographer, and writer with PBS, Christian Science Monitor, and independently as Riley Films. She has also worked in education, as well as museums and libraries on Cape Cod. She has a BA in anthropology from Wellesley College and has been gardening for more than thirty years, blueberry farming for fifteen years.

Conservation specialist Stephen Spear has worked for the United States Department of Agriculture (USDA) for more than thirty years in the Natural Resources Conservation Service (NRCS). He holds a BA in English from the University of Utah and a BS in plant and soil science from the University of Massachusetts, Amherst. He has been a farmer and gardener for more than forty years. The couple lives on Cape Cod, at the farmstead in East Dennis and at their pond house down the road in Brewster, with their two dogs and cats.